ABERRATIONS OF BRITISH BUTTERFLIES

Dedicated to E. V. R. and H. G. M. M.

ABERRATIONS OF BRITISH BUTTERFLIES

By

A. D. A. RUSSWURM

Forty colour plates
by the author

1978

E. W. CLASSEY LTD

E. W. Classey Ltd
Park Road
Faringdon
Oxon. SN7 7DR

ISBN 0–86096–002–1

Printed by Butler & Tanner Ltd
Frome and London

Paper Huntsman Blade Cartridge 135 gsm

Set in Monophoto Baskerville

CONTENTS

FOREWORD

I consider it a great privilege to have been invited to contribute a foreword to this volume in which Donald Russwurm has once more been able to display the excellence of his skill in depicting in colour many aberrant forms of our butterflies. In the 40 beautiful plates he has drawn and painted well over 300 figures of these insects, in many instances showing both the upperside and the underside which differ so remarkably from the normal markings. The origin and date of capture of each individual is most concisely given too. What is of especial interest and importance is that many of these abnormal forms are of recent record and apparently very few have been reproduced before in any published work.

In 1938 the late F. W. Frohawk produced his *Varieties of British Butterflies* embracing a fine selection up to that date of the more extreme aberrations of our near 60 species which breed in the British Isles. It is therefore a most happy occasion that the style of this great artist has been followed by the superb brushwork of the present author so as to provide, in this book, such a worthy successor to the earlier one. How Nature with her many vagaries produces these extraordinary aberrant forms is most problematical, though they seem to occur more frequently when the population of any particular species is in great profusion. Many theories have been put forward as to the cause of their occurrence. However this may be, the publication of this fine volume is a most welcome addition to our knowledge of these abnormal insects. The author is indeed much to be congratulated for their very faithful reproduction in colour so that any reader can well appreciate their natural appearance. The whole work should commend itself to all interested in our butterfly fauna and also to all those who are interested in and lovers of Natural History in general.

THE BARON CHARLES DE WORMS, M.A., Ph.D., F.R.I.C., F.L.S., F.R.E.S.

Charles de Worms

LIST OF PLATES

INTRODUCTION

Many spectacular aberrations have been captured since the publication in 1938 of F. W. Frohawk's fine book *Varieties of British Butterflies.* The present work is an attempt to continue on the same lines and to describe and illustrate some of the magnificent aberrations captured since that date. The coloured plates have been assembled in a style similar to that adopted by Mr Frohawk, who in the author's opinion is still the most admired of butterfly artists.

In the last fifty years the study of aberration has made much progress and almost developed into a science in itself. The complex work of naming the more outstanding aberrations has also progressed but is made more difficult by the many transitional forms that occur, so that it must sometimes be left to the individual to decide in which category a particular specimen should be placed.

On the mainland of Europe and throughout the Palaearctic Region, stretching from Europe across Asia to Japan, some of our familiar British species are represented by subspecies or geographical races. It is to these local races that continental entomologists pay more attention rather than to the individual aberrations so highly prized by British collectors. In this country, where our native species are restricted in numbers, a few years only are required for collectors to take their short series of "typical" specimens; after that the hunt for varieties and aberrations commences. This often requires long hours in the field and much physical endurance, sometimes patrolling the same ground throughout the day. Every major aberration captured is a prize for the collector and can give much pleasure over the years. So often the aberrant form is the more beautiful and in view of the rarity of aberrations and consequent small number of specimens taken, this is one form of collecting that can do no harm to the continuance of the species, nor should it give offence to the conservationist.

There is no doubt that out-of-season temperatures and other vicissitudes of our variable climate are the major factors that play a part in producing aberrations. There is evidence of this in that, due to the record-breaking high temperatures during the summer of 1976, a remarkable number of aberrations were recorded that year. This was particularly noticeable with the Silver-washed Fritillary, of which a number of extreme aberrations were captured within a restricted area. Through the kindness of the entomologists concerned I have been able to figure six of these remarkable insects. The effect of changes of temperature were described by F. W. Frohawk who wrote: "There are several natural forces that influence the production of aberrations. As far as we know, the chief cause is climatic. Great changes of temperature, sudden extremes of light rays penetrating through foliage during stages of development, shock caused by concussion, a sudden noise may bring about aberration. Such disturbances are most effective when the creature is undergoing a change from larva to pupa, at which time the whole of the tissues are soft." Much research has been carried out by entomologists who have subjected fresh pupae to extremes of temperature, and aberrations similar to those found in nature have resulted. Varieties produced by this method are regarded as artificial products and have no place in this work.

It is remarkable how in some colonies a species may remain consistent and show little or no variation whilst in another locality this same species may be very variable, even adopting a different form. This is attributed to a difference in climatic conditions or geographical isolation and these races are usually termed subspecies. When an individual shows extreme variation from the type the term aberration is used. A factor which must be mentioned is that variation is inherited; in this case the particular variety will be found in the same colony of insects year after year. An example of this phenomenon is ab. *fowleri* of the Chalk-hill Blue where various Dorset colonies are famous for the fact that this beautiful variety may be found in small numbers every year. Dr E. B. Ford in his books *Butterflies* and *Moths* sets out a detailed account of the genetics of lepidoptera to which interested entomologists are referred.

The majority of those species that are prone to variation will be described when they are dealt with individually in the text, but before proceeding with this it may be helpful to discuss the different forms of aberration that occur. First, *melanism* can be dealt with briefly. It shows an increase in the black markings and is most prevalent in the Fritillaries, where the black spots and bars may be enlarged or confluent in extreme cases the black pigment may spread over all the wings obscuring the normal markings or the ground colour may be suffused with black scaling giving a dusky appearance. A number of these melanic specimens are illustrated.

Albinism affects the black markings only. Due to a reduction or absence of black pigment they are reduced to mauvish grey or even whitish. Occasionally the Fritillaries and some of the *Satyridae* (Browns) may have their ground colour reduced to creamy white—the author has on occasion seen these colour forms mistakenly labelled albinos. In the true albino the ground colour remains unchanged, the black markings only being affected.

Dimorphism may be divided into three classes. A species may be described as being dimorphic when it appears in more than one colour form. We have two outstanding examples in this country. The female of the Clouded Yellow (*Colias croceus* Geoffroy) appears with the rich orange-yellow ground colour reduced to creamy white or white. More rarely intermediate forms occur, ranging through various shades of primrose yellow to the normal. Our other example is the Silver-washed Fritillary (*Argynnis paphia* Linnaeus). An average of about one in ten of the females appears in the beautiful green-grey and silver *valesina* form. Some scientists believe that this form is a reversion to an earlier type when in primeval times butterflies were much more soberly coloured, the bright colours we are familiar with developing later. (Intermediate forms are very rare, the author having captured one example only; this was taken in the New Forest in 1963 with the normal female colouring but with areas of *valesina* grey in the centre of each hindwing.) Seasonal dimorphism occurs in some families when there is more than one brood during the summer; an example of this can be seen in some of the Whites or *Pieridae*: the black markings in the spring are often but not always dusted over with grey scales. Lastly there is sexual dimorphism where sometimes male and female differ to such an extent that in earlier times they were classified as distinct species. The most striking examples occur in the Blues or (*Lycaenidae*) where in some species the females are smoky brown and the males blue.

Gynandromorphism shows indications of both sexes apparent in one specimen. In the past these specimens were described as hermaphrodites but this term is incorrect. The most striking examples occur in the Blues where the blackish-brown female may be streaked with the blue of the male, or in more extreme cases may be completely halved, male on one side and female on the other. These are referred to as bilateral gynandromorphs; the antennae, thorax and abdomen can be affected in the same way. Some striking examples are illustrated in the Adonis and Chalk-hill Blues, and the Scotch Argus. Intersex specimens have a similar appearance but are genuinely of one sex; this occurs most often in the Silver-studded Blue where a female may have a dusting of the blue scales of the male or even areas of blue on the wings.

Teratological. Deformed or misshapen specimens and other abnormalities come under this heading; these are not attractive to the eye and spoil the appearance of any cabinet drawer or coloured plate. They are not illustrated here and we must assume they belong to a more scientific approach than is claimed by this book. Some pathological specimens are shown where areas of the wings may be bleached or discoloured; sometimes these areas are symmetrical and the effect is not displeasing.

Homoeosis. This form of aberration, although generally very rare, occurs occasionally in members of the *Satyridae* and *Pieridae* where markings and areas of ground colour of the forewings may be reproduced on the hindwings. Some examples captured by the author of both the Meadow Brown and the Gatekeeper show the bright orange of the underside of the forewings reproduced in streaks and splashed across the hindwings, and in several specimens the black apical spot has also been reproduced.

Somatic mosaic. Another very rare form which, although of one sex, somewhat resembles a gynandromorph, one side differing from the other, giving a halved or mixed effect. A specimen with the markings aberrant on one side only may also be classed in this category.

The selection of aberrations figured has been made by the author with the aim of showing as wide a range of variation as possible within the space available. There must be omissions that will disappoint some entomologists and this is regretted. Some of the rare migrant species are omitted, as their appearance in this country is so erratic that little of note has been recorded in the way of variation. Finally this book is intended as a tribute to those lepidopterists whose names have become famous for their work in the field in the past. Our museums have been enriched by the fine aberrations, the data labels of which bear their names.

Many of my entomological friends have given assistance and much useful advice with this work. I would like to make special mention of Mr and Mrs A. J. Pickles for reading through the manuscript and making a number of useful and helpful suggestions. To Mr A. S. Harmer for providing many facilities for the work and for the loan of a number of aberrations from his collection now figured on the plates. To Mr and Mrs R. W. Watson for the unrestricted use of their fine collection, now part of the British Museum (Natural History), and for their unlimited hospitality while painting the large number of specimens figured from their collection. To the staff at the Dept of Entomology for their cooperation when figuring specimens from the National Collection and especially to Mr D. S. Fletcher. My thanks also to many other entomologist friends for the loan of their specimens; these include the following: I. G. Farwell, M. Hale, Dr J. W. O. Holmes, G. E. Hyde, S. F. Imber, Major-General C. G. Lipscomb, C.B., D.S.O., Dr C. J. Luckens, H. G. M. Middleton, H. Phelps, R. C. Revels, Paul Smart, T. W. Tolman, R. S. Tubbs, O.B.E., L. D. Young.

DESCRIPTION OF ABERRATIONS

HESPERIIDAE

PLATE 1 — The Skippers

The Hesperiidae or Skippers as they are known to many entomologists are with two exceptions fairly well distributed over the southern half of England, and all except two of our eight species are common. The family is distributed throughout the world and in the tropics the species assume larger and more colourful forms far removed from the small quietly coloured representatives found in this country. Their rapid flight as they dart from flower to flower has earned them their nickname of Skippers. With their short stocky bodies they more closely resemble small moths than butterflies. The Dingy Skipper, in its normal resting position with the forewings folded back over the hindwings, can easily be mistaken for a small Noctuid moth. By some biologists this family is regarded as a link between the butterflies and moths.

Carterocephalus palaemon Pallas The Chequered Skipper

This is the rarest of the group, at one time thought to be restricted to a few localities in the Midland counties. In recent years it has become even more scarce, having died out from a number of its haunts. In 1942 a colony was discovered in western Inverness-shire where it is still flourishing. Further colonies have since been found in western Scotland. It is a woodland species with a rapid flight, settling frequently on flower-heads and with a strong preference for the flowers of bugle, this plant being in flower at the time when the butterfly is in flight, i.e. from the latter half of May until mid-June. Abroad it has a very wide range extending across Europe, Asia to Japan. It is also found in parts of North America.

Variation consists mainly of enlargement or reduction of the orange spots. They may be enlarged to form a confluent band on the forewings, ab. *scabellata* Leeds; at the other extreme the forewings may be so black that the yellow spots if visible are reduced to mere lines, on the hindwings the marginal spots are almost invisible, referred to as ab. *nigra* Derenne. A more extreme form where only the central spot remains on the hindwings is named ab. *extrema* Dioszeghy. An albino form with all the dark markings replaced by pale brownish grey is ab. *albinotica* Goodson.

Thymelicus sylvestris Poda The Small Skipper

A species that is common in many parts of England Wales but becomes scarce in the north; in Scotland it is very rare. Abroad it is found throughout Europe and North Africa. It is fond of wild flowers and can be found in almost all kinds of uncultivated ground. Although single brooded it is in flight over a fairly long period, being on the wing from the end of June until the end of August.

Variation is mostly concerned with the ground colour, which varies from the normal fulvous through shades of straw colour to almost white. An extreme example of the white form is illustrated and is a recent capture from Dorset; it is named ab. *pallida* Tutt. A less extreme form with the ground colour straw-yellow is ab. *intermedia* Frohawk. A rather dark colouring with a blackish suffusion on the hindwings is named ab. *obscura* Tutt; with all the wings darkish brown, ab. *suffusa* Tutt.

Thymelicus lineola Ochsenheimer The Essex Skipper

This species is restricted to the southern and south-eastern counties of England and closely resembles the Small Skipper, so much so that it can only be safely distinguished in the field by the black tips to the underside of the antennae. Its haunts are similar and it is on the wing during the same period. Variation mostly concerns the ground colour. A yellowish-white colouring is ab. *pallida* Mosley, a deep chocolate brown is ab. *brunnea* Tutt. A female upperside may have the dark border on the hindwings enlarged to occupy one fifth of the wing and is named ab. *marginatus* Picard.

Thymelicus acteon Rottemburg The Lulworth Skipper

There is very little variation in this species, although a fine example of the female is illustrated giving a general effect of black and white colouring and lacking all the golden-olive; this has been named ab. *alba* Bolton. Another form with all the wings pale straw is named ab. *pallida* Frohawk. Last, a golden-brown colouring with the paler markings barely visible is ab. *clara* Tutt.

A restricted area along our south coast from Swanage through Dorset to Devon appears to be the extreme northern limit of its range. Abroad it occurs throughout central and southern Europe and north-west Africa.

Hesperia comma Linnaeus The Silver-spotted Skipper

This species occurs on some of the chalk downs of southern England. It is usually fairly common where it occurs but it is extremely local. It has not been recorded from Scotland or Ireland. It has a wide range abroad, extending eastwards through Europe to Asia and Japan, and is also widely distributed throughout North America where there are a large number of named forms. It flies during the first week in August and is on the wing for about three weeks. It has a rapid flight, probably the fastest of the group and is almost impossible to follow on the wing. There is some variation in the ground colour particularly in the female. A bright fulvous colouring with only slight traces of the cream spotting and with the usual dark border almost obsolete has been named ab. *clara* Tutt. A heavily suffused form is ab. *suffusa* Tutt with all the darker colouring replaced by silvery grey, ab. *albescens* Oberthür. On the underside of the hindwings the silver spots of the outer row are reduced to the three top ones with an indistinct fourth, the lower ones being obsolete is ab. *dupuyi* Oberthür. At the other extreme the white spots may be enlarged and united to form a large white blotch in the centre of the hindwing, this is ab. *juncta* Tutt.

Ochlodes venata Bremer & Grey The Large Skipper

A handsome species common throughout England and Wales and has a wide range abroad, extending eastwards through Europe to China and Japan. It is also found in Africa. Variation is similar to the other Skippers and mostly concerns the ground colour, which varies from straw-yellow, ab. *intermedia* Frohawk, to bright fulvous with the marginal areas particularly dark, ab. *clara* Tutt. With the ground colour intense dark brown and with the spotting small and indistinct it is named ab. *obscura* Tutt. This form is not uncommon in the New Forest where the author has taken a series of both sexes over the years. Another even darker form with all the wings dark brown has been named ab. *fuscus* by Frohawk.

The butterfly is on the wing from early in June until the middle of July or in some years even later. It flies in the rides and on the outskirts of woods and in other kinds of grassy waste ground.

Erynnis tages Linnaeus The Dingy Skipper

It has already been remarked how closely this active little butterfly resembles a small Noctuid moth when at rest. It has a rapid flight, settling most often on the ground where it will sun itself with wings expanded.

It is not attracted by flowers. The butterfly is found throughout England and Wales wherever there is rough uncultivated ground; in Scotland and Ireland it is more local. Abroad it extends as far as western Asia. In favourable seasons it appears in flight as early as April but in more normal seasons it flies during May and June.

There is not a great deal of variation; however the ground colour has some named forms and the transverse bands on the upperside of the forewings vary somewhat. A drab brown specimen with the transverse bands well developed has been named ab. *transversa* Tutt. A blackish fuscous colouring with a strongly grey median area is ab. *suffusa-variegata* Tutt. A drab ground colour with ill-developed transverse markings is referred to as ab. *alcoides* Tutt; with all the wings of a uniform brownish black and no transverse bands visible, ab. *unicolor* Freyer.

Pyrgus malvae Linnaeus The Grizzled Skipper

A species that is abundant but does not occur as far north as the Dingy Skipper. The two species appear on the wing at the same time and are often seen in flight together. Variation concerns the white spotting mainly on the forewings. A rather common form has the two white spots on the inner margin of the forewings united to form a white line, the other spots remaining normal—this is known as ab. *scabellata* Reverdin. Another rather rare form has the white spots on the forewing confluent on both upper and underside forming a large white central area. This striking aberration is known as ab. *taras* Bergsträsser. Although generally rare, in a few localities this variety has been recorded year after year, while in other areas, where the species is abundant, it may never occur. In another aberration intermediate between the "type" and ab. *taras* the white spots above the inner margin of the forewings are united into an oblong blotch and the discoidal spot is also joined to the next spot in another oblong blotch. The hindwings of both these aberrations have a row of white marginal spots and one in the centre of the wing. A form that has the ground colour brown instead of black has been named ab. *brunnea* Tutt.

PAPILIONIDAE

PLATE 2

Papilio machaon Linnaeus, ssp. *britannicus* Seitz The Swallowtail

This beautiful species was widely distributed throughout this country in the past but is now restricted to a few breeding grounds in the fen districts of Norfolk, Suffolk, and Cambridgeshire. No explanation has been offered for its choice of habitat in this country since on the Continent it is equally at home in woods, fields, and on mountains up to an altitude of 5000 ft. It is represented in this country by a subspecies known as *britannicus* Seitz. This differs from the continental form which is more heavily marked with yellow. The English subspecies has heavier black veins and the black band on the outer area of the forewing is wider and increases in width towards the lower part of the wing whereas in the continental form the band remains parallel. These points of difference are important as continental specimens of ssp. *bigeneratus* occasionally cross the Channel and have been known to start small colonies in our southern counties but none of these has prospered for long.

There is almost no minor variation in this beautiful insect but melanic specimens sometimes occur in which all the wings are heavily suffused with black scaling with the original markings just showing through; this is known as ab. *obscura* Frohawk. A more extreme form has all the wings brownish black with no trace of the markings; this is ab. *niger* Heyne. Entomologists have from time to time captured these black specimens and through inbreeding have kept the strain going. In all these melanic forms the blue scaling remains unimpaired. Specimens with the ground colour dark yellow or orange yellow are sometimes found in collections. These are not always genuine and may be the result of chemical action; however this form does occasionally occur when the species is bred in captivity and has been named ab. *burdigalensis* Trimoulet. At the other extreme an almost white ground colour is referred to as ab. *pallida* Tutt. A handsome aberration that has much blue scaling is known as ab. *eminens* Schultz. The marginal band on the forewings is dusted with bright blue scales instead of yellow; on the hindwings the band is blue with no black scaling visible.

PIERIDAE

PLATE 3

Colias croceus Geoffroy The Clouded Yellow

Variation in this beautiful insect is mainly concerned with the ground colour and in the yellow markings in the black borders of the female; the male remains fairly constant. The species is dimorphic in the female which appears in two distinct colour forms. The form *helice* Hübner has the rich yellow ground colour replaced by creamy white; between this and the normal occur every intermediate shade, although in the author's experience the cream-coloured form is the more often met with. In the historic invasion of this species in 1947 the *helice* form actually outnumbered the typical female, at least in the areas worked by the author. In the most extreme examples of *helice* Hübner the ground colour may be reduced to dead white. This form is named ab. *alba* Lempke and may also have the orange spot in the centre of the hindwings replaced by yellow so that the name ab. *flavomaculata* Braun may be added, or the spot may be white similar to the ground colour, ab. *albissima* Ragusa. The yellow spots in the borders of the female may be reduced or completely absent (referred to as ab. *pseudomas* Cockerell); the name ab. *obsoleta* Tutt also seems to refer to this form. The specimen illustrated on Plate 3 shows this aberration in the form *helice*. Another very beautiful aberration may be mentioned; this has the wings shot with purple or blue or violet. These varying shades are placed under the name ab. *purpurascens* Cockerell.

The home of this lovely insect is in southern Europe and North Africa. It is a notorious migrant and entomologists in the last century who knew little or nothing about migration were greatly puzzled by its erratic appearance. In some seasons very few specimens may be observed although it is rarely completely absent from our shores. About once in every six or seven years they may arrive in swarms, spreading over the whole country and reaching as far north as the Shetlands and the Orkneys. These large migrations have unfortunately become much less frequent in recent years.

Arrivals in the spring deposit their eggs and produce a brood later in the summer, but the brood is not really at home in our climate and the British-bred specimens are often undersized and usually easily distinguished from fresh arrivals from overseas.

Colias hyale Linnaeus The Pale Clouded Yellow

Very little variation has been recorded in this species, and for this reason it is not illustrated. With a home further east and a somewhat less powerful flight it is consequently a less frequent visitor to this country. Our winter climate is not suitable for it in any stage so that its appearance here is solely due to migration. Arrivals in May and June can produce a small brood in August and September. There has however been less evidence of migration during this century and the species is now regarded as something of a rarity. The most likely haunts are fields of clover and lucerne in southern England, usually not far from the coastal areas.

Records show that the black markings on the upperside are somewhat variable. On the forewings the yellow spots may be enlarged and run together to form a yellow band so that the black area is divided by a yellow band; this is known as ab. *flavoradiata* or ab. *flavofasciata* Lempke. An albino specimen with the black markings in mauvish grey has been recorded and named ab. *albinotica* Goodson. A mixed gynandromorph was bred by Mr N. A. Watkins from Kentish parents in 1948. This specimen is male on both wings on the left-hand side. On the right side the forewing is female, the right hindwing mostly female but with areas of male scaling. It is somewhat unusual that the wings and antenna on the female side are smaller than on the male side. The specimen is now in the Watson collection of the British Museum (Natural History).

Colias australis Verity The Berger's Clouded Yellow

This species is very similar to the Pale Clouded Yellow and it was not until 1947 that it was identified as a separate species. The larvae differ considerably as also does the foodplant, being horse-shoe vetch in the Berger's Clouded Yellow. This explains why the butterfly is more often found flying over rough chalky ground and hillsides where the foodplant grows. The species is much the rarest of our three Clouded Yellows, possibly due to being less migratory in its habits. Any variation would probably be similar to the other members of the *Colias* group.

Plate 4

Gonepteryx-rhamni Linnaeus The Brimstone

Except perhaps for the Small Tortoiseshell this butterfly lives longer than any other of our species in the adult stage. It emerges from the chrysalis towards the end of July and is often still in flight when the next generation appears on the wing the following year. It is remarkable how the male retains the brilliant yellow colouring even after twelve months as an adult. The Brimstone is one of the first butterflies to appear in the spring after hibernation and sometimes as early as February on a mild sunny day it will make a bright spot of colour as it flies briskly over the winter landscape.

Due to the absence of markings there is little scope for variation. Very rarely the orange spot on the forewings may be enlarged and may spread over the centre of the wing so that the specimen somewhat resembles the south European species *G. cleopatra* Linnaeus. Gynandromorphs sometimes occur, either completely halved or less rarely as mixed gynandromorphs, showing a mixture of the colouring of both sexes, as shown in the specimen illustrated on Plate 4. Occasionally specimens occur blotched with red or brown or brownish red, but the extreme example illustrated with one side almost completely red and with the colour spreading over the other side is believed to be unique; no similar specimen is known to the author. This and other specimens showing varying shades of red or brown can be placed under the name ab. *rubescens* Gillmer. Equally remarkable is the specimen figured in the centre of the plate with all the wings heavily suffused with black scaling giving the effect of olive-green colouring, referred to as ab. *nigrescens* Hechler.

Plate 5

Leptidea sinapis Linnaeus The Wood White

A very fragile and delicate species with a slow and rather heavy flight which gives the impression that the wing muscles are hardly strong enough to carry the very slender body. It is mainly a woodland species although there is one colony firmly established along a stretch of cliff on the coast of Devon. It flies slowly around the woodland rides, settling on the lower foliage. In dull weather it rests on the underside of the leaves.

There is very little variation but the beautifully marked subspecies from Ireland known as ssp. *juvernica* Williams is illustrated although it cannot be classified as an aberration. In this form the ground colour on the underside is yellow with the greenish markings clearly defined. Albinistic aberration with the markings pale sandy brown is known as ab. *brunneomaculata* Stauder. In recent years this aberration has been recorded a number of times from the Midland counties. A form with the dark markings on the underside absent is named ab. *diniensis* Boisduval. Specimens entirely devoid of markings on both upper and underside are referred to as ab. *erysimi* Borkhausen.

Although the butterfly has become extinct in a number of areas including the New Forest it is still found in many parts of England and Wales.

It is generally distributed in Ireland although local. It appears on the wing in May and June and in favourable years there is a partial second brood in July and August. Specimens from the spring brood are more heavily marked and are referred to as form *lathyri* Hübner.

PLATE 5

Pieris brassicae Linnaeus The Large White

This resident species is reinforced by migration from the Continent and although there is fluctuation in numbers from season to season it is usually sufficiently abundant for the larvae to cause much damage among garden produce. The butterflies arrive from the Continent in swarms; eggs are then deposited on all plants of the cabbage family, including Nasturtium, which is much favoured. Fortunately parasites take a heavy toll of the larvae so that a certain balance is maintained and the destruction is limited.

There is little variation, but there is evidence of seasonal dimorphism as the black markings in the spring specimens may be dusted over with grey scales, although this is not always so. The two black spots on the forewings of the female may be linked together by a dusting of black scales, which form is known as ab. *fasciata* Kieffer. An albinistic form has been bred in captivity and is referred to as ab. *albinensis* Gardiner. A very beautiful form has been bred in recent years and is known as ab. *coerulea* Gardiner. On the underside the normal cream or yellow is replaced by pale blue, the upperside is thinly scaled, giving a pure white appearance instead of the normal creamy white. Only two previous examples of this beautiful aberration were known to the author. There is also a record of a sulphur-yellow male referred to as ab. *flava* Kane.

PLATE 5

Pieris rapae Linnaeus The Small White

Closely related to the Large White, this species is even more abundant and causes great damage among cultivated vegetables. Its habits are much the same as that of the previous species. It is double brooded and there is a partial third brood so that it is in flight almost throughout the summer. There is seasonal dimorphism as in the Large White but in this species it is more clearly defined, the black markings being more intense in the summer brood.

There are not many named varieties, and of these the more important are as follows. A yellowish form found more often in Ireland is ab. *flava* Ter Haar. A light brownish or coppery ground colour is ab. *brunneoflavida* Stauder. Joining together of the two black spots on the forewings of the female by a dusting of black scales is referred to as ab. *fasciata* Tutt. Very rarely the upper of the two black spots may be joined to the apical mark by a similar dusting of black scales; this is known as ab. *conjuncta* Mezger. A male specimen is figured showing complete absence of black markings on the forewings, referred to as ab. *immaculata* de Selys. The hindwings of this specimen show a heavy suffusion of black scaling on the underside.

PLATE 6

Pieris napi Linnaeus The Green-veined White

Unlike its two near relatives, the Large and Small Whites, this species can never be regarded as a garden pest, its foodplants—being garlic mustard and horseradish—belong more to the country lanes and hedges than to our gardens. The species is double brooded and the spring brood shows the same seasonal differences as other members of the family, i.e. dusting over of the black markings on the upperside with grey scales in the spring brood; the veins on the underside however are much more heavily defined in this brood.

The specimens illustrated show the result of much interbreeding by entomologists. The rich lemon-yellow form, known as ab. *sulphurea* Schoyen, from parts of Ireland being paired with banded specimens having the two black spots on the forewings enlarged to form an irregular band, ab. *fasciata* Kautz. A male and female of this double aberration are shown in Figs. 1 and 2. The same aberration is shown on Fig. 3 in the albino form. These albino specimens are referred to as ab. *pallidus* Frohawk and may occur in both white and yellow races. In Scotland and parts of Ireland females may sometimes appear with an ochreous or even pinkish tinge, ab. *flavescens* Tutt; two examples of this form are shown. In some of the larger collections may be found gynandromorphic specimens both halved and mixed; some of these may show a mixture of the white and yellow colour forms.

PLATE 6

Anthocharis cardamines Linnaeus The Orange-tip

This is one of the first butterflies to appear in the spring and in early seasons it has been noted on the wing in April. It has a brisk flight and is gaily coloured but apart from that does not have a very interesting history, as nine months of its life-cycle is passed in the pupal stage.

The species is a striking example of sexual dimorphism, as the orange patch of the male does not appear in the female. It may be for this reason that mixed gynandromorphs are not so rarely seen in this species as in most others, being more easily detected on the wing. Two examples are illustrated. Although there are completely halved specimens on record they are much more rarely met with. Albino specimens have occurred having no trace of black scaling and are known as ab. *lasthenia* Mill. The black discoidal spot is usually larger in the female than in the male and has been recorded as being completely absent in the male although very rarely. This aberration has been named ab. *immaculata* Pabst. The discoidal spot may be rounded in shape as in ab. *crassipuncta* Mezq., or shaped like a crescent, named ab. *parvipuncta* Turati. The remaining aberrations concern the orange patch in the male. Reduced to lemon yellow it is referred to as ab. *aureoflavescens* Cockerell. A deep brownish colour is named ab. *buschmanni* Müller. Dwarfed specimens of no more than 33 mm are not uncommon.

LYCAENIDAE

PLATE 7 The Hairstreaks

The Hairstreaks are a group of small but attractive butterflies numbering no more than five species in this country. There is little variation amongst them and the majority of the few aberrations recorded are shown on Plate 7. There is usually a thin white line across the hindwings which end in a short tail.

Callophrys rubi Linnaeus The Green Hairstreak

Most often seen is the Green Hairstreak. It is well distributed in this country and has a wide range abroad. It is usually found where the food-plants of broom and furze grow, such as the outskirts of woods, open downs and hillsides. The only variation of the upperside recorded is a slight bleaching or ochreous colouring on the outer half of the wings. The beautiful green underside may occasionally be brownish and is then known as ab. *brunnea* Tutt. In some northern counties the small white spots on the underside may be enlarged to form an irregular band mainly across the hindwings; the author has met with this form in Cumbria. Sometimes the white spots may be absent, as in ab. *caecus* Geoffroy.

Thecla betulae Linnaeus The Brown Hairstreak

This is a very shy species and is rarely seen in the adult stage. The larvae however can be beaten from the bushes in a blackthorn thicket and most collectors obtain their short series in this way. The butterfly is found in most of the southern and midland counties of England and Wales; abroad its range extends across central and northern Europe to eastern China. The rich orange underside is extremely beautiful and shows the usual white hairlike streak across all wings. A very rare aberration, shown on

Fig. 9, has this streak absent from all wings; it is known as ab. *steinbuhleri* Hoffman. Another rare underside aberration, ab. *unistrigata* Schultz, has been observed to appear most often among specimens bred in captivity. The orange markings at the base of the hindwings have been known to join up and form a band, and the orange patch in the centre of the forewings may be reduced to ochreous yellow in colour.

The larvae may be beaten in June and when bred in captivity the butterflies may emerge during July. In the wild, August, September and early October are the months that the butterfly is on the wing.

Quercusia quercus Linnaeus The Purple Hairstreak

This species is common in most of the oak woods in England and Wales, particularly in the south. There are a few localities in western Scotland and in Ireland it is scarce and very local. Abroad the species is found throughout Europe but does not extend beyond Asia Minor. The butterfly fluctuates somewhat in numbers, being very abundant in certain years although in others it can be quite scarce. Most of the day is passed flying around the top of oak trees but it descends to the ground for the night and is best looked for during the morning, when it can often be seen sunning itself on bracken with wings expanded.

There is some variation in the upperside. The dark purplish grey of the male may be replaced by rich purple, known as ab. *violacia* Niepelt. The purple area on the forewings of the female may be blue instead of purple, referred to as ab. *caerulescens* Lempke. Another aberration of the female has a short series of yellowish-orange marks on the forewings as illustrated, and is named ab. *flavimaculatus* Lienard (=*bellus* Gerhardt). Two very extreme underside aberrations are illustrated and named.

Strymonidia w-album Knoch The White-letter Hairstreak

A very rare and striking aberration of this species was captured by Major-General C. G. Lipscomb in Wiltshire and is shown on Plate 7, Fig. 3. It is named ab. *albovirgata* Tutt. There is very little other variation recorded except for the orange-red band on the underside of the hindwings which may occasionally be enlarged, when it is known as ab. *rufextensa* Goodson. A specimen of this aberration is illustrated in *South's British Butterflies*, Plate 11, Fig. 3. The species is very local but may abound in favoured localities where there are plenty of elm trees. Occasionally stray specimens may be seen far removed from the breeding ground. It is mostly a southern species and does not extend further north than Yorkshire.

Strymonidia pruni Linnaeus The Black Hairstreak

A very local species, being confined to a few of the midland counties, mainly Oxfordshire and Northamptonshire. The orange markings on the upperside of the female may sometimes be enlarged to form a band on both fore and hindwings; this has been named ab. *ptorsas* Hufnagel. An underside aberration has the white streak enlarged to form a band on all wings similar to ab. *albovirgata* Tutt of the White-letter Hairstreak and is named ab. *albofasciata* Tutt in this species. This aberration is also figured in T. G. Howarth's *South's British Butterflies* on Plate 11, Fig. 6.

Lycaena phlaeas Linnaeus The Small Copper

Although firmly established in Britain this small brightly coloured species is more at home in a sunnier climate than ours. A cool and rainy summer can seriously deplete its numbers, whilst in a dry and warm season it will flourish and produce a substantial third brood in the autumn. It is found throughout the British Isles and is much attracted by flowers. Favourite haunts are downs, hillsides and flowery fields, in fact almost wherever there are wild flowers.

There are many named aberrations of this species and these can be divided into several groups; there is variation in the black spotting on the forewings, in the shades of copper on both fore and hindwings, and in the copper band on the hindwings. In the second brood the copper on the forewings may be heavily suffused with dark scales. This form is more often met with during a hot summer and is similar to the summer form in some of the warmer countries of southern Europe; it is named ab. *fuscae* Robson. The copper ground colour of the forewings may be reduced to cream or straw colour in ab. *cuprinus* Peyerimhoff, more rarely it may be silvery white as in ab. *schmidtii* Gerhardt. Spotting on the forewings may be absent except for the discoidal pair and this is referable as ab. *bipunctata* Tutt. When the spots are reduced in size it is known as ab. *parvipuncta* Tutt. A specimen of this form combined with the light *cuprinus* colouring is illustrated. The black spots may be elongated so that they join up with the discoidal pair and this is referred to as ab. *extensa-conjuncta* Tutt.

The hindwings are subject to variation in the copper band. It may be completely absent as in ab. *obsoleta* Tutt, or replaced by a few streaks of copper running up the veins; this is named ab. *radiata* Tutt. A fairly common form has a row of blue spots inside the copper band and is known as ab. *caeruleopunctata* Ruhl. In an extremely rare form this blue spotting on the hindwings may be replaced by a copper colouring. The specimen shown on Fig. 19 is a new aberration bred by Dr J. W. O. Holmes and has been named ab. *coeruleo-fasciata* Holmes. I am indebted to Dr Holmes for the following notes: "One of four similar aberrations bred from a series of 37 out of an F2 of 138. These were bred from a strongly marked blue-spotted wild female. An F3 also of 37 were bred true to type." A magnificent example of ab. *berviniensis* Smith, now in the British Museum (Natural History), was captured by Capt. A. P. Gainsford in October 1969. It is in a class by itself and does not belong to any of the groups mentioned above.

Cupido minimus Fuessly The Small Blue

This is the smallest of our butterflies and although very abundant in its chosen haunts it is very local. Favourite localities are rough ground where there are plenty of wild flowers and where the foodplant of kidney vetch grows in abundance. It appears to have a preference for coastal districts in the chalk and limestone areas but it is not restricted to these localities. Although distributed all over the British Isles it is far more plentiful in the south. The butterfly emerges towards the end of May and flies throughout June. In warm dry summers there is a partial second brood in August. The butterfly has a rapid flight and flies near the ground. It settles frequently on grasses and rests with wings expanded in the sun.

The female is uniform smoky brown, the male blackish, with a dusting of steely blue scales. The density of the blue scaling is variable as also is the colouring. It may be greenish in tint, when it is named ab. *viridescens* Tutt, or violet as in ab. *violascens* Tutt. A pale grey form of the male is referred to as ab. *pallida* Tutt. The spotting on the underside varies considerably, and specimens with all spotting absent except the discoidals are not uncommon; this aberration has been named ab. *caeca* Courv. When the spotting is reduced in size it is referred to as ab. *parvipuncta* Tutt, and when enlarged as ab. *magnipuncta* Tutt. Very rarely the spots may be elongated into small dashes or streaks; the specimen is then termed ab. *striata* Tutt. There is a fine series of this very rare aberration in the British Museum (Natural History). Very small specimens have been named ab. *minor* Tutt. At the other extreme, when larger than normal it is referred to as ab. *major* Tutt.

PLATE 8

Celastrina argiolus Linnaeus The Holly Blue

The habits of this pretty species more closely resemble those of the Hairstreaks than the Blues. It is often seen flying around the upper branches of holly trees except that single specimens will sometimes stray far from the breeding ground when they may be seen flying briskly across gardens. The wandering habit of this species may account for the fact that it will occasionally extend its habitat and turn up in an area where it had not been recorded previously.

The female provides an example of seasonal dimorphism, those of the partial second brood being more heavily bordered with black around the forewings. Another unusual feature of this interesting species is that each brood has a different foodplant. Eggs for the spring brood are laid on the flower-buds of holly while those of the summer brood are laid in a similar way on ivy.

The butterfly has a very wide range. It is locally common over the southern half of England and Wales but north of the Midlands and in Ireland it is rather local. The British race is known as subspecies *britanna* Verity. Abroad it extends across Europe through northern Asia to China and Japan, then across to central and North America where according to Holland there are many named forms. It is also found in North Africa. Considering its wide range not many aberrations have been recorded. The light grey form of the male, ab. *grisescens* Eisner, is illustrated (Fig. 1). This form also occurs in the female where the blue is replaced by grey. Other named colour forms mostly occur in the female and several deserve mention. Where the upperside is mauve with broad black marginal borders it is known as ab. *lilacina-lata* Tutt. A violet white female is named ab. *pallida* Tutt. Reduced spotting on the underside tending towards *obsolescens* is ab. *paucipuncta* Courv.; complete absence of spotting except the discoidals is ab. *obsoleta* Tutt. At the other extreme the topmost spot on the forewings and almost all the spots on the hindwings are elongated into black streaks and is referred to as ab. *subtus-radiata* Oberthur. A magnificent example is illustrated (Fig. 3). Some gynandromorphs have been recorded but these are very rare.

Plate 8

Plebejus argus Linnaeus The Silver-studded Blue

In this country there are four distinct subspecies of the Silver-studded Blue and although this work is mainly concerned with aberrations, a brief description of the characteristics of the four races is considered necessary. The nominate form, ssp. *argus argus* Linnaeus, occurs on heaths, moorlands, and heather-clad commons in the south. This is locally common in Surrey and Dorset and very abundant in the New Forest. In the chalk and limestone districts, mainly in Kent, a race of large specimens with whitish undersides occurs and is named ssp. *cretaceus* Tutt. Figs. 8 and 9 show some of the aberrations of this subspecies and the light coloured underside is clearly indicated. From North Wales a race of dwarf specimens occurs in the limestone districts along the coast of Caernarvonshire; the females have more blue on the upperside than those from further south. This subspecies is named *caernensis* Thompson. Lastly a magnificent race once found in the mosses of Witherslack was destroyed by a disastrous fire in 1941. The specimens were of large size and the very blue females greatly prized by collectors. The name given to this subspecies is *masseyi* Tutt.

There is some variation in the colour on the upperside of the male.

A lilac tint is named ab. *lilacina* Tutt, at the other extreme an almost black male with only a slight blue sheen is known as ab. *obscura* Grund. The black marginal border is variable and may be very narrow when it is named ab. *augusta marginata* Tutt, with the black border greatly enlarged it is referable as ab. *late-marginata* Tutt. The ground colour of the underside can be dark grey or even blackish as in ab. *infraobscura* Lempke. In the author's experience this form occurs more frequently in the male.

There is an aberration of the female where the lunules on the upperside are white; this is known as ab. *albopunctata* Galvagni & Preissecker. On the underside the lunules are much given to variation in colour. An example with the lunules dark brown, ab. *fuscescens* Lempke, is illustrated on Fig. 15. Another form with deep red lunules is ab. *rufescens* Tutt, and with lemon-yellow lunules is ab. *flavescens* Tutt. The spotting on the underside varies in much the same way as other members of this family. Several examples are illustrated where on all four wings the submedian spots are connected with the marginal chevrons with black streaks; this is known as ab. *juncta* Tutt. A more extreme form has a black streak from the base to the discoidal spot and this is sometimes joined to the submedians, and is known as ab. *extrema* Courv. With the spotting more or less obsolete it is referred to as ab. *privata* Courv.; with all the spots absent ab. *caeca* Grund. It is interesting to compare the gynandromorph shown on Fig. 6 with the intersex Fig. 7. The intersex is a female with the blue scaling of the male dusted over the brown ground colour of the female giving a mauve effect compared with the clear blue of the male section in the gynandromorph.

Plate 8

Aricia agestis Denis & Schiffermüller   The Brown Argus

This is an active and sun-loving insect, common over the southern half of England and Wales; northwards, it extends as far as the Peak District where it is replaced by the Northern Brown Argus, recently classified as a distinct species. The Brown Argus is double brooded, first appearing in May and June, the second brood on the wing in July and August. It is found on rough uncultivated ground, sheltered slopes and hollows on flowery hillsides. In dull weather and towards evening the butterflies rest, usually in small groups, on flower-heads and the heads of grasses.

Variation concerns the orange lunules, the colour ranging from cream as in ab. *pallidior* Oberthür, to completely white in ab. *graafi* Ver-Huell. There is also some variation in the size of the lunules; in the males they may be reduced in size or partially absent. The spotting on the underside varies in the same way as other members of the family. Complete absence of spotting is referred to as ab. *obsoleta* Tutt. Some examples of *radiata* forms are illustrated wherein the submedian spots are joined to the marginal chevrons; this has been named ab. *subtus-radiata* Oberthür and is extremely rare in this species. Occasionally the discoidal spot on the forewings is edged with white in a form known as ab. *snelleni* Ter Haar. This is reminiscent of the next species where the discoidals are typically completely white.

Plate 8

Aricia artaxerxes Fabricius The Northern Brown Argus

Previously classified as a subspecies of the English *A. agestis* it was not until 1967 that it was proved conclusively that two separate species occur in Britain. *Aricia artaxerxes* is found throughout Scotland and extends as far south as the Durham area across to the Lake District. These specimens from northern England now belong to the subspecies *salmacis* Stephens and are believed to interbreed with the English *A. agestis* which would be at the northern limit of its range.

The Scottish *A. artaxerxes* is easily distinguished from the English species by the white discoidal spot on the forewings; more rarely it appears on all four wings when it is known as ab. *quadripuncta* Tutt. On the underside, the black pupils are either greatly reduced or completely absent from the white spots. The ground colour is generally darker on both upper and underside than in *agestis*. This is more in evidence in the ssp. *salmacis* Stephens. Two aberrations of this race are shown. Fig. 22 shows the complete absence of the orange lunules on the upperside. A characteristic of this species is that the lunules are generally smaller than the English *agestis* and when absent altogether they are referred to as ab. *unicolor* Lempke. Partial absence of spotting as in Fig. 23 is named ab. *obsoleta* Tutt.

Plate 8

Hamearis lucina Linnaeus The Duke of Burgundy Fritillary

This small woodland butterfly is the only British representative of a family centred in South and Central America. It is somewhat misnamed as a Fritillary as it has no connection with the family *Nymphalidae*. It does however bear a superficial resemblance to the Heath Fritillary, which no doubt accounts for its early English name. Although usually a woodland species it prefers the sunny rides and glades, settling on the lower foliage rather than flowers. It has also been known to breed on the open downland and was at one time common in a Dorset locality which was famous among entomologists.

Three aberrations are figured including an albino specimen, but apart from these there is not much variation in this species. The amount of black on the upperside is somewhat variable. With the black markings reduced it is referred to as ab. *gracilens* Derenne, an increase in black is ab. *obscura* Aigner. In the female the black spots along the outer margins of all wings may be absent as in ab. *obsoleta* Tutt.

The Duke of Burgundy is one of the first butterflies to appear in the spring and is in flight from the middle of May throughout June. There are a few colonies in Lancashire and Yorkshire, but generally it is restricted to central and southern England.

Plate 9

Polyommatus icarus Rottemburg The Common Blue

Although common throughout the whole of the British Isles, this species never appears in quite the same profusion as occurs with the Chalk-hill and Adonis Blues in their most favoured localities. The Common Blue is double brooded in the south of England and in favourable summers it is likely that there is a third brood, as it can then be seen in flight throughout most of the summer. In the north of England and in Scotland there is one brood only and specimens from these single-brooded races are generally larger than those from the south. It is double brooded in the south of Ireland but in the rest of that country there is but one brood, where it is represented by the beautiful subspecies *mariscolore* Kane. The females of this race produce the largest specimens of all and are extremely handsome with their rich blue ground colour and enlarged lunules of orange or red on all wings.

There is a great deal of variation in both upper and underside and many named varieties. The upperside of the male is mostly concerned with the ground colour. An example with leaden-grey colouring is illustrated (Fig. 1) and is named ab. *livida* Gillmer. Another form of pale lilac-blue is termed ab. *pallida* Tutt. Variation in the females mostly concerns the amount of blue and the arrangement and colour of the lunules. A female with the wings all blue is referable to ab. *caerulea* Fuchs; with the blue restricted to the basal area it is known as ab. *caerulescens* Wheeler. An entirely brown female with complete absence of orange lunules is referable to ab. *fusca* Gillmer. With the blue restricted to the basal area again without a trace of lunules it is ab. *fusca-caerulescens* Tutt. At the other extreme with the orange lunules drawn out into rays towards the discoidal spot it is ab. *rufina* Oberthür. A specimen of this rare and beautiful aberration is figured (Plate 9, Fig. 6). The marginal lunules may be reduced to pale yellow or whitish in ab. *flavescens* Tutt. A form with the upperside very thinly scaled so that the spots from the underside show through is termed ab. *transparens* Tutt. Variation on the underside follows the same pattern as with the other Blues, i.e. linking together the black spots into bars or streaks producing *radiata* and *striata* forms. Complete absence of spotting on the underside is known as ab. *obsoleta* Gillmer.

Gynandromorphic specimens, although rare, are more frequent in this species than with the other Blues and several examples are illustrated. A completely halved or bilateral gynandromorph is shown on Fig. 2, two mixed gynandromorphs on Figs. 3 and 4.

Plates 10 & 11

Lysandra coridon Poda The Chalk-hill Blue

There are more than four hundred named varieties of this beautiful species. They include various colour forms of the male and female upperside; on the underside variation commences with the joining together of two or more of the black spots, then progressing through all stages to the extreme forms of *striata* and *radiata*. Some striking examples are illustrated on Plate 11. It is not surprising that the species is very popular with entomologists as it gives almost unlimited opportunities for studying the many varietal forms. The butterfly is found in suitable localities in the counties of southern England; it is single brooded and in normal seasons appears in flight during the third week in July and continues emerging during the next four or five weeks.

Although still found in abundance in many of its former haunts it has on occasion disappeared from some of the areas where it has previously been very common. Unlike many other species it has remarkable powers of recovery and will sometimes build up gradually to its original numbers. The reason for this periodical decline can only be conjectured. Possibly an increase in the parasites may upset the balance, or over-grazing by cattle may restrict its food supply and these alternatives appear to be the most likely causes. The most favoured localities are flowery hillsides and open downland in districts of chalk and limestone. When working this species in the past, the author has recollections of sun-drenched landscapes with fields of ripe corn stretching away until lost in the August heat haze. With our variable climate this cannot however be described as typical August weather.

The colouring of the male upperside varies considerably and some of the named forms should be mentioned. A pale and more silvery blue than the typical is named ab. *pallidula* Tutt., a distinct greenish tint as ab. *viridescens* Tutt, and a darker and more intense green is known as ab. *ultraviridescens* Tutt. A beautiful sky-blue form is named ab. *caeruleo* Tutt. The blue lustre may be absent altogether and replaced by pale slate grey when it is known as ab. *grisea* Tutt, and this form leads through varying shades of grey to very dark grey referable as ab. *pulla* B. & L. The most extreme example is almost black and may appear to be thinly scaled and is known as ab. *plumbescens* B. & L. This colour also appears in the female, giving a uniform blackish appearance. In these very dark specimens the underside is darkish and heavily suffused. The black margin of the forewings is variable in width. A slight widening is named ab. *inframarginata* Tutt; when the black scaling extends across the wings to the discoidal it is known as ab. *marginata* Tutt, when it extends beyond that to the base of the wings is referred to as ab. *meluina* Tutt. In these extreme forms the hindwings may also be heavily bordered with black then the name ab. *latiora* B. & L. may be added. The beautiful ab. *fowleri* South with the black borders replaced by white occurs along the south coast from Portland to Swanage, although it has also been recorded from some inland localities. Some male examples are illustrated including the rare combination of ab. *fowleri* South with the grey ground colour of ab. *grisea* Tutt. In a more extreme form the white border extends around the hindwings obscuring the black dots and is referred to as ab. *ultrafowleri* B. & L. Further examples illustrated show the result of selected pairings between ab. *fowleri* South and ab. *tithonus* Meigen and also with ab. *semisyngrapha* Tutt, being the work of two very skilled and experienced entomologists. Variations of the female upperside include a completely blue female, distinguished from the male only by the blackish-brown thorax and abdomen retaining the normal female colouring. This has been known to entomologists as ab. *syngrapha* Keferst; however the earlier name ab. *tithonus* Meigen now takes precedence. In ab. *semisyngrapha* Tutt the hindwings are blue but there is less blue on the forewing. The sections between the two lower veins must be blue right to the border. If less than this it is referred to as ab. *infrasemisyngrapha* Tutt. Plate 10, Fig. 1 shows a male with the black borders replaced by cinnamon brown; this colouring also occurs in the female which may in addition have a rosy tint. It is known as ab. *cinnameus* Tutt. The underside of both sexes contains almost every form of variation recorded in this family from a complete absence of spotting, ab. *caeca* Courv., or partial absence, ab. *obsoleta* Tutt. Some rare and striking aberrations of the underside are illustrated on Plate 11, showing the spots joined into black streaks and classified under ab. *radiata* Courv. and *striata* Tutt forms. A whitish ground colour in both sexes is referred to as ab. *alba* Tutt. An aberration not uncommon in the female but less frequent in the male has the basal and submedian spot on the forewings joined, forming a straight line, and is named ab. *i-nigrum* Tutt. This line may be curved either upwards or downwards and is then referable as ab. *arcuata* Courv. Both bilateral and mixed gynandromorphs have been recorded in this species, in view of the sharp contrast in colouring between the sexes they are conspicuous when in flight, which may account for the fact they seem to occur more frequently in this species. Sometimes there may be streaks or larger areas of bright blue on the wings of the female, and in some examples there may be traces of black towards the borders. This is termed ab. *inaequalis* Tutt. A striking example is illustrated on Plate 10, Fig. 10.

Interbreeding between two different species is almost unknown in this country, but Mr T. G. Howarth writes in his *South's British Butterflies* that "male specimens that are intermediate in colour between *bellargus* and *coridon* are thought to be natural hybrids between these two species, as apparently they only occur in localities where these two butterflies are present". This rare butterfly is known as *polonus* Zeller.

We have touched briefly on the more important and spectacular aberrations but for those who wish to make a more extensive study of this most rewarding species a complete and comprehensive work was published in 1938 entitled *A Monograph of the British Aberrations of the Chalk-hill Blue Butterfly* by P. M. Bright and H. A. Leeds. At the time of writing this work is out of print but if a copy can be obtained the work is strongly recommended.

PLATE 12

Lysandra bellargus Rottemburg The Adonis Blue

The haunts of this species are similar to those of the Chalk-hill Blue but being more local in its distribution they are not always found together. It is still abundant in some of its localities but has died out in a number of areas where previously it had been very abundant. It is restricted to southern England and is not found north of Bedfordshire and Oxfordshire. One is most likely to find it on the chalk and limestone hills of Kent, Surrey and Sussex. The species is double brooded, first appearing towards the end of May and throughout June, then again in August and September.

Variation is similar to that found in the Chalk-hill Blue and other members of the family. A number of colour forms of the male have been named; these include a pale lavender colouring which is referred to as ab. *pallida* Mosley; a greenish tint, ab. *viridescens* Tutt. A dark blue-black colouring is *totonigra* Lipscomb; this form is illustrated on Fig. 2 and is believed to be unique. When strongly tinged with purple or violet it is known as ab. *purpurascens* Tutt. A blue-black colouring is referred to as ab. *nigra* Cockerell, a dark leaden blue as ab. *suffusa* Tutt. Very rarely the colouring may be half-way between the blue of *coridon* and that of *bellargus* and such forms are believed to be hybrids between the two species. It is named ab. *polonus* Zeller (see also under *Coridon*).

The blue female, with all its intermediate forms, corresponds with *tithonus* and *semisyngrapha* of *L. coridon*. With the blue scaling extending over the fore and hindwings obliterating the ground colour except along the costa and outer margins it is named ab. *ceronus* Esper; in this aberration the orange lunules are present on all four wings. Ab. *semiceronus* Tutt is similar but the orange lunules are present on the hindwings only. In ab. *caerulescens-lunulata* Tutt the forewings are blue from the base to beyond the discoidal point, the hindwings are blue from the centre outwards and the orange lunules are well developed. The females of the first brood mostly have a dusting of blue scales at the base of the wings then through all stages leading up to the all-blue ab. *ceronus* Esper. The author has observed these blue females only in the spring and early summer specimens, those of the late summer brood being of a uniform dull smoky black.

On the underside, with two submarginal spots joined to the discoidal it is named ab. *parallela* Courv., and is similar to the *L. coridon* aberration of the same name. In ab. *striata* Tutt all the wings are heavily streaked and rayed with black. With all spotting absent it is referred to as ab. *krodeli* Gillmer, with the spotting very small ab. *parvipuncta* Tutt. Figs. 3 and 4 show gynandromorphs but these are very rare.

Plate 13

Maculinea arion Linnaeus The Large Blue

This species has for some time been in danger of extinction in this country and is now protected by Act of Parliament 1975. It follows therefore that any aberrations figured must belong to the past. The butterfly has disappeared from a number of localities where it was abundant during the last century and is now restricted to a few localities in Devon and Cornwall, and somewhat doubtfully in the Cotswolds where specimens differ from those from Cornwall, being sprinkled with black scaling giving a melanic appearance. This race is known as ab. *cotswoldensis* le Chamberlain and resembles the form found in Brittany.

The somewhat unusual life-history wherein the larvae become carnivorous after the third and final moult and complete their growth feeding on ant larvae is fully described in the textbooks. The first emergence, consisting mainly of males, takes place during the last week in June. The females appear during the second and third weeks of July and by this time the males are showing signs of wear. A similar instance occurs in the Silver-washed Fritillary, the females emerging about eight days after the first males have appeared. The most favoured localities are rough hillsides and valleys with an abundance of wild thyme, heather and gorse, usually facing south along the coastal areas.

The markings on the underside remain remarkably consistent but on the upperside there is considerable variation in the spotting. It ranges from being completely absent except for the discoidals, ab. *arthurus* Oberthür. When partly obsolete it is referred to as ab. *alconides* Aurivillius. When three spots are all that remain of the normal row on the forewings it is named ab. *supra-impunctata* (given in Seitz) Oberthür. This reduction in spotting occurs more frequently in the male; see Figs. 7 and 8. The female varies in the other direction, the spotting being enlarged into wedges and streaks. This is named ab. *insubrica* Vorbrodt. Some examples of this aberration are shown on Figs. 3, 4 and 6.

Some colour forms of the upperside have been named. A dull light grey ground colour is named ab. *grisea* Courv. A pale washed-out appearance is referred to as ab. *pallida* le Chamberlin. A dwarf form has occurred in all localities and is known as ab. *occidentalis* le Chamberlin.

NYMPHALIDAE

PLATE 14

Ladoga camilla Linnaeus The White Admiral

Variation in this fine insect is restricted to melanism, i.e. suffusion or dusting over of the white markings with black scaling. The markings may be partially blacked out with the pattern still visible, this has been named ab. *obliterae* Robson & Gardner and is shown on Fig. 1 on the plate. More rarely the markings may be completely absent; it is then referable as ab. *nigrina* Weymer. It will be seen from the illustrations that the white markings on the underside are correspondingly reduced or absent. The melanic specimens seem to occur in cycles and they may be absent for a number of years, then in a single year they will occur in most of the woods in southern England. The author took three ab. *obliterae* in July 1956 then had to wait nine years before capturing two out of three observed in 1965. Some extreme examples were recorded during the very hot summer of 1976 and further similar aberrations were taken during 1977. It would appear from observations that the completely black form ab. *nigrina* Weymer is usually among the later emergences.

The butterfly is a lovely sight when freshly emerged, with its beautifully marked underside and graceful sailing flight, as it glides up and down the sunlit rides in the forest. The brambles however soon take a heavy toll on the beautiful wings, but long after they have been reduced to tatters the graceful flight remains unimpaired. Fortunately this species has extended its range and increased in numbers during the last fifty years.

Plates 15 & 16

Apatura iris Linnaeus The Purple Emperor

With its large size and striking colouring this fine insect is considered the most beautiful of our native species, and might well have been adopted as our national butterfly. The Japanese gave this title to the most admired of all their butterflies, the magnificent *Sasakia charonda* Hewitson, which in their own country they call the Great Purple.

Due to its shy and retiring habits it has always been regarded as a great prize by collectors; it is more often seen than captured. Much of its day is passed at rest on a chosen leaf at the top of some lofty oak tree. Being somewhat pugnacious in habit it will attack and fight off any intruder. Its depraved taste for carrion and other decaying matter will sometimes prove its undoing when it can be lured to the ground, and when imbibing fall an easy prey to the net. The long periods at rest are broken about once every hour when it will take a short flight, sailing round the upper branches but always returning to its chosen resting place, often to the same leaf.

Variation follows the same pattern as that found in the White Admiral, and is mainly in the degree of dusting out of the white markings on both upper and underside. Extreme forms with the white markings reduced to one or two spots are placed under the name ab. *lugenda* Cabeau. Another form with the white markings reduced or partly missing has been known as ab. *semi-iole* Frohawk; now however the earlier name ab. *iolata* Cabeau is preferred. When the white markings are completely absent it is referred to as ab. *iole* Schiff. This aberration is extremely rare and very few examples are known to exist.

In view of its retiring habits it is difficult to assess the actual numbers breeding in this country but in recent years it appears to have become more numerous and it now occurs in most of the larger oak woods in the southern half of England.

Plates 17 & 18

Aglais urticae Linnaeus The Small Tortoiseshell

There is a great deal of variation both in the ground colour and in the black markings of this common but very pretty species. Entomologists breed the larvae in large numbers in the hope of obtaining aberrations. Being gregarious the larvae are easy enough to find, as they live in large masses in untidy grey webs spread out over stinging nettles.

Extreme aberrations are less rare in this species than most and some outstanding examples are illustrated. A number of these were captured in the wild, including the two specimens shown on Plate 19. The specimen figured above was taken by the author when it was flying around a sunny clearing in the Pignel enclosure of the New Forest, the one below being captured near Ringwood, Hants, at an earlier date. A similar example coupled with pale whitish ground colour is shown on Plate 18. The rich orange colouring varies through shades of pinkish yellow to cream or even whitish, the latter being known as ab. *lutea* Raynor. Sometimes the costal spot is joined to the black spot on the inner margin by a dusting of black scales, or more rarely may be completely blacked in to form a solid band and is known as ab. *pseudoconnexa* Cabeau.

Occasionally one isolated aberration may occur in a brood where all the remainder are typical specimens. The author had one instance of this when a brood of typical specimens was bred from a web of larvae taken from Hod Hill in Dorset. All were released except that one pupa was overlooked. Three days later a fine example of ab. *semiichnusoides* Pronin emerged and was found at rest on the floor of the breeding cage (Plate 19, Fig. 1). This late emergence can be explained by the fact that extreme aberrations are often weak and do not always succeed in breaking open the pupa casing and so perish without emerging. In this case emergence had been delayed by several days.

Plate 19

Vanessa cardui Linnaeus The Painted Lady

A lovely migrant species but not a very regular visitor to this country as some years it is absent altogether. The butterfly has no hibernating stage and as our winter is totally unsuited to it in any of its stages of development, its appearance in this country is dependent on its migratory habits. It is a notorious wanderer and it is reported that there is not a country in the world where its appearance has not been recorded at some time or other. Vast swarms of the butterflies have been met with at sea many hundreds of miles from land.

Fig. 1 on Plate 20 shows ab. *varini* Meilhan. This is a form of the one extreme aberration of this butterfly; there are also modifications or transitional forms. A development of ab. *varini* has the black markings forming the median band on the forewings absent between the discoidal cell and the inner margin, and the large white costal blotch is heavily obscured. Ab. *rogeri* Meilhan is a further development of ab. *varini* Meilhan, the discal band being absent on the forewings and the submarginal spots on the hindwings, although reduced in ab. *varini*, are represented by white ocelli instead of black.

There are some minor aberrations and these include ab. *melanosa* Cabeau where the black markings on the forewings are very heavy with the fulvous area much restricted. The hindwings are similar with the submarginal spots enlarged and confluent. A rather common form has the hindwings suffused and smoky and is referable to ab. *semisuffusa* Cockerell. Another form has blue centres to the black spots on the hindwings and is known as ab. *ocellata* Rebel. A dwarf form only half the normal size is called ab. *carduelina* Alpheraky.

Plate 19

Vanessa atalanta Linnaeus The Red Admiral

Another beautiful migrant species but a much more frequent visitor than the Painted Lady. It may fluctuate in numbers from year to year but it never seems to be entirely absent from this country during the summer. Arrivals from the south in early spring deposit their eggs on nettle and have no difficulty in reaching the adult stage during the summer months, producing the butterflies that we are familiar with on our buddleias in August. It is also often seen feasting on fallen rotting fruit. Their numbers by this time have been augmented by fresh arrivals from the south but like the Painted Lady it is considered most unlikely that it survives the winter, the earliest butterflies seen on the wing the following spring being fresh migrants.

The specimen figured of ab. *klemensiewiczi* Schille is the main aberration and remains remarkably consistent, but the red bands are subject to much variation in colour. However these varieties are very rare and are seldom seen in private collections. The majority of these colour forms have been named and the principal ones are as follows. When the red bands are completely white it is known as ab. *millierei* Cabeau. A straw-yellow form is referable to ab. *flavescens* Fritsch. When orange it is ab. *eos* Fritsch. Finally a brownish-ochre colour possibly due to some black suffusion is ab. *ochrobrunnea* Fritsch. The bands on the forewings may be heavily suffused and almost blacked out—it is ab. *umbrosa* Fischer.

Plate 19

Inachis io Linnaeus The Peacock

A beautiful and hardy native whose velvety-black larvae live spread out in large colonies on the beds of stinging nettle and must be familiar to most people. Best known among the aberrations is what is known in the vernacular as "The Blind Peacock", with the beautiful blue "eye" markings absent. All that is visible is the dull grey ground colour in which they would normally be set. This aberration is very rare in nature but occasionally turns up when the species is bred in captivity. There are many intermediate forms leading up to this aberration and one of these is figured. The completely "blind" Peacock is named ab. *belisaria* Oberthür.

There are a number of lesser aberrations and some of these have been named. An important one with the ground colour greyish violet and usually with a yellowish tinge to the centre of the forewings gives a general impression of a transparent effect. This is believed to be due to thin or faulty scaling. This form has been named by a number of authors, such as ab. *fulva* Oudemans, ab. *transparens* Beuret, and ab. *violascens* Rebel. Another colour variety that is not uncommon has the ground colour dark chocolate brown and is referable to ab. *brunnea* Reuss. The author has bred this form a number of times; usually a complete web of larvae will produce this colouring.

Plate 20

Nymphalis polychloros Linnaeus The Large Tortoiseshell

Only one major aberration has been recorded of this scarce insect and two examples are illustrated on Plate 21. The magnificent specimen figured below was bred from a web of about eighty larvae found on sallow near Lady Cross in the New Forest in 1902. Six of these aberrations emerged (two deformed). It is named ab. *testudo* Esper. Other variation is comparatively slight, i.e. the ground colour may have a mauvish tinge and there may occasionally be blackish areas formed by a dusting of black scales on the forewings below the twin black spots.

The species has at times become so scarce that it was believed to have died out altogether in this country but somehow it has managed to survive and on a few occasions has actually increased in numbers. The larvae are gregarious and fall an easy prey to parasites; in this way whole colonies have been destroyed. When fully grown the larvae are reported to release their hold on the foodplant, which is usually elm, and fall to the ground; they then wander away in search of a suitable place to pupate, usually a nearby fence or a tree trunk. The author has found larvae in this way on the Continent but has had no experience of them in this country.

The butterfly is not often seen in the perfect state, as shortly after its emergence from the pupa in late July or August it goes into hibernation and does not reappear until early spring. This probably accounts for the comparatively fresh condition of the hibernated specimens. It may occasionally be seen sunning itself in lanes on the outskirts of woods either on the ground or on tree trunks.

Plate 21

Polygonia c-album Linnaeus The Comma

This beautiful and interesting species was rather scarce and local during the last century, it is pleasant to report that during recent years it has greatly increased in numbers and is now fairly common over the southern half of England and Wales, extending as far north as the Wash across to the Mersey.

The species is double brooded, the first appearing in June and July, the second being on the wing from late August onwards. These later butterflies are more often seen than the earlier ones, being found in company with the Red Admirals when feeding on fallen fruit during September. It was in this way that the author captured the aberration shown on Fig. 3 of the plate. This is one of the two major aberrations that occur in this species; the other is shown on Figs. 1 and 2 and is named ab. *reichenstettensis* Fettig.

There is considerable variation in the ground colour although extremes are very rare. The dark chocolate-brown form is known as ab. *neole* Oliver. In the other direction it ranges through shades of buff and cream to the completely white form ab. *dilutus* Frohawk. The white "comma" mark on the underside of the hindwings varies in shape; it is sometimes joined across to resemble the letter "O" when it is named ab. *o-album* Tutt. It may also be shaped as a letter "G" and is known as ab. *g-album* Tutt. It also varies in intensity, being sometimes reduced to a thin hairlike streak, ab. *extincta* Rebel. This latter is less rare than the other forms, the author having taken several specimens. Very rarely the white "comma" may be completely absent when it is named ab. *c-extinctum* Gillmer, but this form is recorded as having occurred only when the specimen has been bred in captivity.

A different colour form occurs among the butterflies of the first brood and is known as form *hutchinsoni* Robson. The colouring is much lighter and more ochreous on both upper and underside. The wings are also less variegated, particularly in the female.

Plate 22

Boloria (*Clossiana*) *selene* Denis & Schiffermüller The Small Pearl-bordered Fritillary

This small but very pretty species is closely related to the Pearl-bordered Fritillary, the next on the list. They are much alike but are most easily distinguished by the underside. It appears on the wing early in June when the latter is becoming worn and nearly over. Both species follow much the same form of variation so that in the case of an extreme aberration some consideration may be necessary to determine which species is concerned.

The Small Pearl-bordered appears to be more prone to variation than our other Fritillaries and certain localities have been known to produce major aberrations year after year. These may range from a completely white ground colour (see Fig. 1) through shades of cream and buff to a completely melanic specimen. Continuing with the colour forms a light brownish-yellow ground colour is named ab. *flavescens* Lempke. This is considerably paler than the normal. An almost black form is referred to as ab. *vela* Motschulsky. An albino specimen with normal ground colour but with all the black markings in dull grey is ab. *decolorata* Oberthur. The five specimens figured show a wide range of variation but there are a great many intermediate forms leading up to those illustrated.

The butterfly is a woodland species but not essentially so, as it may also be found in damp meadows provided there is a supply of the food-plant, dog violet, near at hand. It can also be found on hillsides and railway cuttings. It is well distributed throughout Scotland but in England and Wales seems more plentiful in the south and west. It is rather more local than the Pearl-bordered but common enough where it is found. It does not occur in Ireland.

Plate 23

Boloria (*Clossiana*) *euphrosyne* Linnaeus The Pearl-bordered Fritillary

This small Fritillary closely resembles the preceding species on the upper-side but is easily distinguished on the underside, being warmer in colour with a distinct orange tinge and the markings are less contrasted. It makes an early appearance, being on the wing from the middle of May, but in early seasons it has been seen in flight from the middle of April. This may account for its early English name of the "April Fritillary". It has a swift and graceful flight but is not always easy to follow on the wing as it flies briskly over dead bracken, which it matches closely in colour.

Variation is similar to the Small Pearl-bordered Fritillary but occurs rather less frequently. In addition to the four aberrations figured there are many intermediate or transitional forms which have to be classified under those already named. Variation in the ground colour has been graded into three stages. The normal rich fulvous orange may be reduced to pale yellow-ochre when it is referred to as ab. *albinea* Lambillion; next a pale ochreous colouring is named ab. *stramineus* Frohawk. The most extreme is a completely white specimen but still with the normal black markings, named ab. *pallida* Spuler. A pale orange-yellow ground colour but becoming lighter towards the borders and even whitish on the hind-wings is known as ab. *luteus* Frohawk. An albino form with all the black markings replaced by silvery-grey but with the normal ground colour has been named *albamaculata* Goodson. Sometimes the black spots on all wings may be joined to form a zigzag line; this may vary in width and is referred to as ab. *transversa* Vorbrodt. Another aberration described and named *tatrica* by Aigner has the median spots in the forewings extended into a broad black band with a large black blotch in the cell and with a further one below. The basal half of the hindwings is also black. An almost black specimen is illustrated and named ab. *nigricans* by Oberthür. The underside of the hindwings of this aberration is beautifully rayed with silver and is also illustrated. At the other extreme the black markings on the upperside may be reduced to a minimum and often more or less obsolete at the margins, is referred to as ab. *pauca* Tutt. An underside variation has the silver spots on the border of the hindwings replaced by leaden grey and is known as ab. *plumbea* Cockayne.

Plate 24

Argynnis (*Fabriciana*) *adippe* Linnaeus The High Brown Fritillary

This large and handsome Fritillary has a powerful and graceful flight. It is essentially a woodland species and is fond of feasting on brambles and thistles on the outskirts of woods. In dull weather it rests in the trees. It is well distributed over the southern half of England but its range does not extend into Scotland or Ireland.

Variation is not so rare and three extreme aberrations are illustrated. Many transitional forms between these and the normal have been recorded. There are also some fine albino specimens in existence where the black markings are replaced by whitish grey. Some of these have been illustrated in the past. Another remarkable aberration has the silver spots absent from the underside; they are indicated by a lighter shade of buff than the surrounding ground colour. This aberration, known as ab. *cleodoxa* Ochsenheimer, is very rare in this country but according to one authority is more often met with in southern Europe, especially in Greece and Sicily. Very dark females with the upperside smoky have been named ab. *berolinensis* Reuss. A still darker form named ab. *bronzus* Frohawk has the forewings black except for some spots of the normal orange ground colour in the cell. The hindwings are dusky except for the basal half which is of the normal orange. With the base of the forewings suffused and with the black spots enlarged and united it is known as ab. *suffusa* Tutt.

Plate 25

Argynnis (*Mesoacidalia*) *aglaja* Linnaeus The Dark Green Fritillary

In the author's opinion the female of this fine species with its rich bronze green lustre on the upperside is the most beautiful of our larger Fritillaries. It has a very powerful flight and is more often seen flying at speed over the open downland, so cannot be regarded as a woodland species in the same way as our other Fritillaries. It is fond of feasting on thistles in fields adjoining woods when it becomes completely engrossed and may then be examined at leisure.

The very wide range extends throughout Scotland as far north as the Orkneys. In these northern districts it is represented by a subspecies named *scotica* Watkins. The females of this race are often heavily suffused with black scaling. Occasionally similar specimens are met with in the south. Other forms of variation include changes in the ground colour and these occur more often in the male. When reduced to creamy white it is known as ab. *albescens* Verity. In ab. *charlotta* Haworth the silver spots at the base of the hindwings on the underside are joined and form several blocks. In this aberration the black markings on the upperside are usually confluent. Melanic aberrations occur, although rarely; a magnificent black female ab. *wimani* Holmgren is illustrated on the plate. An albino specimen which is similar to the albinos of the other Fritillaries has the black markings replaced by silvery grey and is named ab. *albomaculata* Rebel.

Plates 26, 27 & 28

Argynnis paphia Linnaeus The Silver-washed Fritillary

Aberrations of this large and handsome Fritillary are spectacular, as might be expected. Records show that they occur at intervals of approximately thirty years, very little of note being taken in between. Looking back, the most notable years were 1881, 1918/1919, 1941/1942, and 1944. Then after another lean period they appeared again in 1976 when most of the fine aberrations figured in this work were captured.

As mentioned in the introduction the female is dimorphic, appearing in two distinct colour forms, one with the normal fulvous ground colour of the Fritillaries, the other referred to as form *valesina* Esper with a beautiful mixture of grey, bronze-green and silver. This form does not occur in the male except in gynandromorphs (see Plate 26, Fig. 1). There is very little evidence of colouring intermediate between these two forms but one female specimen captured by the author in 1963 is of the normal colouring, but with areas of the *valesina* grey in the centre of each hindwing. Like several other specimens of this form of variation noted by the author there are whitish patches on all four wings.

Variation consists mostly of enlargement and joining together of the black spots to form elongated or eye-shaped markings, hence the name ab. *ocellata* Frings. for this aberration. In extreme examples of melanism the black scaling may cover almost all the wings; it is then named ab. *nigricans* Cosmovici. When the form *valesina* produces a melanic aberration it usually takes this form and is named ab. *nigrizina* Frohawk. There are very few examples of ab. *ocellata* in the *valesina* colouring, although a fine example is illustrated on Plate 27, Fig. 3. White spots sometimes occur on the male; they may be symmetrical but more often they are irregular in their placing. This form of aberration has been named ab. *caroffana* Cabeau.

The life-history of this species deserves some mention, as it must be one of the miracles of nature every time this butterfly reaches the adult stage. The egg is laid on the bark of a tree and the young larva, measuring no more than 2·2 mm in length, on emerging from the egg rests beside the empty egg shell and enters hibernation. Those that survive the winter descend to the ground and wander off in search of the foodplant, dog violet, and so take their first meal after eight months in the larval stage.

Plate 29

Euphydryas aurinia Rottemburg   The Marsh Fritillary

Some of the more interesting points in the life-history of this species are that it remains gregarious throughout its life-cycle. The larvae live closely packed together in a web spun shortly after emergence from the egg. They leave the web to feed during the day and return for the night. This condition continues until they are almost fully grown, when they disperse to pupate. The butterflies on emergence remain within the breeding ground and seldom stray outside.

The species is widely distributed throughout the British Isles in sharply defined colonies so that it is not surprising that it should be represented by a number of named subspecies. The dying out of a number of colonies was probably due to overcrowding and consequent shortage of food. Fresh colonies spring up occasionally to compensate for these losses. The local races have their distinguishing characteristics. In Scotland the sub-species *scotica* Robson has the ground colour generally duller and the black markings heavier. Specimens from the Irish race ssp. *hibernica* Birchall resemble those from the Isle of Wight and also from Cumbria, with the reddish colouring brighter and the black markings more clearly defined; the costa on the forewings is heavily dusted with grey scales. Over the southern and western half of England the species is more evenly coloured, being of a uniform fulvous colouring. There is considerable minor variation and extreme aberrations appear to be less rare in this species than most. Four extreme examples are shown on the plate and a further group appear on Plate 31 of Howarth's *South's British Butterflies*.

Plate 30

Melitaea cinxia Linnaeus   The Glanville Fritillary

This species is more at home in a sunnier climate than ours, its only foothold in this country being on the Isle of Wight, where it breeds mainly along the south coast, although the author has met with single specimens all over the island. Falls of cliff into the sea are not infrequent and whole colonies of larvae are destroyed in this way. Fortunately the species has remarkable powers of recovery—if this were not so it might have been lost to us long ago.

In some seasons the larvae are so numerous that it is difficult to walk along the cliffs without treading on them. The few that collectors take for breeding in the hope of obtaining varieties can do no harm to the species. Possibly the typical specimens they release may account for the small colonies that spring up on the mainland only to die out in a year or two, as the species seems quite unable to establish itself except in its chosen haunts.

On Plate 30, Figs. 1 and 4 show male and female of the most extreme aberration that occurs. Although they differ slightly, particularly on the forewings, they may be placed under the named form ab. *horvathi* Aigner. The hindwings are black and are broken only by a band of fulvous spots before the margin. The dusting of black scaling on the forewings is variable and may be heavy at times. Another aberration has the ground colour white with a yellowish tinge and is named ab. *leucophana* Cabeau. Figs. 2 and 3 show variation in the black markings on the underside of the hindwings. In Fig. 2 they are absent. In the specimen shown on Fig. 3 all the black markings are very broad; the two black wavy lines in the centre of the wing are united into a band of large black spots. This aberration is named *wittei* Geest and occurs occasionally when the species is bred in captivity.

Plate 31

Mellicta athalia Rottemburg The Heath Fritillary

Variation in this species is similar to that found in *M. cinxia* and the other Fritillaries, i.e. melanism or increase in the black pigment sometimes obscuring almost all the orange ground colour. An extreme example of this aberration captured in east Kent is shown on Plate 32, Fig. 4. Previously known as ab. *navarina* de Selys it is now referred to as ab. *cymothoe* Bertolini. The specimen figured immediately above shows a slightly less extreme example from the same colony. These melanic specimens although rare are less so than in our other Fritillaries; this may be due to the extreme abundance of the species in the few restricted areas where it still occurs. Not many years ago the butterfly was so abundant in the east Kent colony that they would fly around one in clouds during a walk along one of the woodland rides; this is not so now. The butterfly will sometimes change its habitat, moving to another area in the wood. This is due to the parasitic nature of the foodplant, which is sometimes choked and destroyed as a result of the surrounding foliage becoming overgrown.

Fig. 1 shows an extreme aberration with the hindwings similar to those of ab. *cymothoe* Bertolini, but the black markings on the forewings are restricted to the base of the wings so it becomes ab. *corythallia* Hübner, previously referred to as ab. *eos* Haworth. In the specimen shown the black markings are again aberrant, being blurred and indistinct, so it becomes a further aberration of ab. *corythallia*. As shown on Fig. 2 the ground colour may be reduced to creamy white and is referable to ab. *latonigena* Spuler.

The species still flourishes in a few colonies in Devon and Cornwall but has disappeared from Sussex. In Essex the chance of its survival is reported to be precarious due to building operations. From the author's observations the specimens from the Devon and Cornish races have shorter and more rounded-shaped wings, giving them a generally smaller appearance than the Kentish specimens.

SATYRIDAE

PLATE 32

Pararge aegeria Linnaeus The Speckled Wood

The British race of this woodland butterfly with its familiar cream-coloured spotting belongs to the subspecies *tircis* Butler=*egerides* Staudinger. This race is found throughout central and northern Europe. The nominate form *aegeria aegeria* occurs in south-western Europe and North Africa and is distinguished by its tawny orange spotting. Specimens are occasionally taken in southern England showing distinct traces of the orange colouring, but these are very rare. One example captured by the author in Hampshire is figured.

Although occasionally found among the foliage of hillsides and cliffs, the butterfly is essentially a woodland species. It appears to shun the sunlight and flies up and down the woodland rides, settling either on the lower foliage or on damp patches on the footpaths. It does not appear to be attracted to flowers. It is common over the southern and western districts of England and Wales, somewhat rare in Scotland, and common all over Ireland.

There is not very much variation. The orange-tinted colour form already mentioned is named ab. *intermediana* by Lempke. Very rarely the cream markings are almost completely obscured by blackish scaling; this is known as ab. *saturatior* Crombrugghe. The eye-spots on the hindwings may be reduced in size, referred to as ab. *parviocellata* Lempke, or they may be without the white pupils on all wings, named *omnicaeca* Lempke, or without white pupils on the hindwings only, ab. *postcaeca* Lempke. The specimen figured with greatly enlarged spotting appears to be unnamed.

Plate 33

Lasiommata megera Linnaeus The Wall

Unlike the previous species the Wall is a sun-loving insect. With a flight that is rapid but usually of short duration it can often be seen flying briskly along hedgerows and settling frequently on bare patches in the sun. It also favours sun-baked walls, and rough ground on the edges of fields. It is widely distributed and often abundant over England, Wales and Ireland. In Scotland it is found mostly in the south-west. There are two broods: the first appears in May and June, the second being on the wing in July and August. The female differs from the male in being larger and without the sexual brand across the forewings.

There is considerable variation in the main apical eye-spot. When considerably enlarged it is known as ab. *anticrassipuncta* Lempke. This aberration occurs almost exclusively in the female. When absent altogether it is ab. *inocellata* Lempke. In another rare form it may be elongated, referred to as ab. *elongata* Lempke. With two white pupils in the main eye-spot instead of the normal one it is named ab. *bipupilla* Mosley. Sometimes in addition to the two white pupils there may be additional spotting, one extra above the main eye-spot and another below. This is referred to as ab. *quadriocellata* Oberthur. With further spotting below those already mentioned it is ab. *croesus* Stauder. A very rare form has the white pupil absent from the eye-spot and is known as ab. *impupillata* Lempke.

There are some variations in the ground colour on the upperside. A light straw colour is named ab. *xanthos* Frohawk. An albino form with the dark markings of a silvery grey but with normal ground colour is ab. *pallescens* Oberthur. A dark suffused colouring on the upperside is named ab. *melania* Oberthur. A specimen with completely white ground colour is known to the author but appears not to have been named. Except for the double pupil in the main eye-spot, which is not uncommon, the remainder of these aberrations are very rare.

Plate 34

Erebia epiphron Knoch The Small Mountain Ringlet

This butterfly is rarely found below an altitude of 1000 ft. It is our only mountain species and in England is restricted to a few localities in the Lake District, where it is represented by the sub-species *memnon* Haworth. In Scotland there are a number of colonies; apart from the well-established race on Ben Nevis in Inverness-shire these are mostly located in the mountains of Perthshire. The Scottish race is known as sub-species *scotica* Cooke and is larger, handsomer and more brightly coloured than the specimens from the north of England. The majority of the specimens figured are of the Scottish race. Apart from the fact that variation is more easily detected in these larger Scottish specimens the author believes that the named aberrations occur more frequently in this race.

In the mild climate of the Lake District the butterfly first appears on the wing about June 20th; in Scotland its emergence is eight or ten days later. Its favourite haunts are damp marshy ground around the mountain streams where it flies fast and straight near the ground, pausing occasionally for a second or two at flower-heads. Sometimes it will be joined by several others when they will set off at speed, so that a long chase is necessary before a capture is made.

This is a most rewarding species to study, as variation is considerable. It concerns mostly the orange-red bands and the black spotting, although bleaching of one or more wings is not uncommon. Fig. 1 shows a male specimen with the orange band reduced to a few separate rings of orange around the black spots. This is named ab. *nelamus* Boisduval. This named aberration covers a fairly wide range among British specimens. Meyer-Dur describes it as "having the band reduced to three or four small suffused spots usually containing small black eye-spots". With the orange bands completely absent it is referred to as ab. *obsoleta* Tutt. This is shown on Fig. 4. A male specimen from the Lake District is shown on Fig. 2 where all the black eye-spots are absent. This is not uncommon among the Scottish specimens and is referred to as ab. *caeca* Vorbrodt. With the spotting very small in the bands it is known as ab. *privata* Turati & Verity. Fig. 5 shows a female with the bands reduced to creamy white, ab. *albinescens* Oberthür. Finally Fig. 6 shows a female with the bands very wide and extending across the forewings to the discoidal cell. This is referred to as ab. *latefasciata* Dioszeghy.

PLATE 34

Erebia aethiops Esper The Scotch Argus

This most attractive species shows a preference for mountainous country being more often found on rough open hillsides. However it also occurs on open moorland, on the outskirts of woods, and has occasionally been found almost at sea level. So, unlike the Small Mountain Ringlet, it cannot be described as an Alpine species. Distribution of the two species is somewhat similar, there being only two colonies of *aethiops* surviving in England, both in Cumbria. A well-known race from the Durham district has become extinct. It is much more widely distributed throughout Scotland, where it is represented by the subspecies *caledonia* Verity. The sub-specific differences are small.

The butterfly is a sun-loving insect, but the author cannot agree with some of the textbooks that state that the moment the sun is obscured the butterfly disappears into the grasses. From personal observations the butterfly will remain in flight for some time, and on dull mornings, provided the weather is mild, it can be seen flying until around noon. This habit is shared by the Ringlet, which it closely resembles in flight. Variable weather and out-of-season temperatures play havoc with the time of appearance of many species, but the Scotch Argus remains remarkably consistent, always emerging during the first few days in August. The male has a slow and lazy flight among the grasses with a liking for bramble blossom. The flight of the female is brief and rapid.

There are two distinct colour forms and these are most noticeable in the female with its lighter colouring. The soft blue-grey bands on the underside of the hindwings of the typical female are replaced by ochreus; this form is known as ab. *ochracea* Mosley. Both of these colour forms are shown among the aberrations illustrated. In the wild they occur in about equal numbers. There is considerable variation in the black spots on the orange-red bands. The normal three spots on the forewing band may be increased to four by the addition of one extra in the centre; this is known as ab. *tetraocellata* Rocci. This aberration is quite common. A further extra spot below making a total of five in ab. *freyeri* Oberthur. This again is not uncommon in some areas. More rarely an extra spot may be placed above the apical spot bringing the total to six, ab. *croesus* Rocci. Occasionally the spots may be unequal in number on opposite sides and is known as ab. *inaequalis* Mosley. When absent altogether from all four wings but with the red bands normal on the forewings and with the band on the hindwings broken into three or four fulvous areas but without spotting, it is known as ab. *inocellata* Owen. With the black spotting normal but with the white pupils absent it is ab. *caeca* Rocci. Again with the black spotting absent and only the white pupils remaining it is referred to as ab. *pupillifera* Kolar. There is some variation in the colouring of the bands; when heavily suffused with black scaling so that they are barely visible against the dark ground colour and the underside similarly suffused it is known as ab. *perfumosa* Dannehl. A specimen of this aberration is shown on Fig. 9. The bands may be pale yellowish, ab. *flavescens* Tutt, or mauvish grey sometimes described as heliotrope, ab. *violacia* Wheeler. Another colour form of the underside of the hindwings of the female has the normal blue-grey bands replaced by whitish grey; this contrasts strongly with the dark bands and is named ab. *leucotaenia* Staudinger (see Fig. 10). This also occurs in the form ab. *ochracea* Mosley where the ochreus bands are more of a creamy tint. A striking aberration where the dark bands are missing altogether and the hindwings of a uniform blue-grey is named *infasciata* Warren. Figs. 8 and 11 show male and female of this aberration. A fine example of the female was captured by Mr H. G. M. Middleton at Arnside in 1971. A similar aberration but in ab. *ochracea* Mosley was taken by the author in the same locality in August 1967. Spotting may be enlarged or elongated but this form of variation does not appear to have been named so must be described as ab. nov. Gynandromorphs as in most species are extremely rare. Two striking examples are shown (Figs. 13–16). One of these in the typical colouring, the other ab. *ochracea* Mosley.

Plate 35

Melanargia galathea Linnaeus The Marbled White

This attractive species is common in most of the counties of southern England but does not extend far into the Midlands; in Scotland and Ireland it is unknown. In distribution it is so local that sometimes a colony may be restricted to a single field. It appears on the wing in July and is a pretty sight as it drifts with its slow lazy flight among the long grasses. It favours rough uncultivated ground on hillsides, meadows, and the outskirts of woods. For no apparent reason a stray specimen may be met with flying far from the nearest breeding ground. In dull weather the butterfly rests on the flower-heads of grasses.

Variation is very rare as the markings remain remarkably consistent. There are some changes in the ground colour but most of the variation concerns the black markings. Mr F. W. Frohawk illustrated and described two specimens showing extremes in both directions. A completely melanic specimen was captured near Rochester in July 1871. A white specimen without a trace of black scaling was taken between Dover and Walmer in July 1843. Some of the more important named aberrations are as follows. With the whitish ground colour replaced by pale citron-yellow, ab. *citrana* Lambillion. With a much deeper yellow which may be a development of the previous aberration, ab. *citrina* Krulikowsky. With the normal black markings replaced by grey, ab. *grisescens* Varin; this form is illustrated on Fig. 1. The black markings towards the outer margin of the forewings may be enlarged to form a black blotch; this is referred to as ab. *valentini* Williams (see Fig. 3). With all the white areas suffused with black scaling but with the markings still clearly visible it is ab. *marconi* Frohawk. With a still heavier suffusion of black scaling and with the markings only just discernible it is referred to as ab. *nigra* Frohawk. With all the marginal spots absent from the underside it is ab. *galene* Ochsenheimer—a specimen is illustrated on Fig. 4. Another extreme aberration with most of the markings in the centre of both fore and hindwings either absent or indicated by a dusting of black scales is known as ab. *mosleyi* Oberthur (see Fig. 2).

Plate 35

Aphantopus hyperantus Linnaeus The Ringlet

The upperside of this species, being almost unrelieved blackish brown, must be the most sombre coloured of all our butterflies, the main interest being concerned with the ringed markings on the underside. It flies in July and is often seen in company with the Marbled White, although it has a slight tendency to be more of a woodland species. It may often be seen flying along the woodland rides, settling frequently on the lower foliage. It is less susceptible to sunlight and on dull mornings will remain in flight until around noon. If the sun has not appeared by then it will gradually disappear into the grasses. Although most plentiful in southern England it is found all over the British Isles.

The majority of the named aberrations concern the ringed markings on the underside. Very rarely they may be absent altogether; this is known as ab. *obsoleta* Tutt. Another aberration that is not uncommon in some districts has the rings absent but the white pupils remain; this is referred to as ab. *arete* Muller. With the underside of the forewings devoid of spots, but with them feebly indicated on the hindwings, it is ab. *caeca* Fuchs. When considerably larger than normal it is known as ab. *crassipuncta* Burkhardt. The rings may be unusually large and pear-shaped with the white pupils extended into short white streaks: it is referred to as ab. *lanceolata* Shipp. Figs. 7 and 8 show both male and female of this rare and handsome aberration. Figs. 5 and 6 immediately above show a rare combination of the drawn-out white pupils of ab. *lanceolata* Shipp and the absence of the outer rings of ab. *arete* Muller. This is the result of breeding from selected pairings between the two aberrations. With the spotting again elongated but less so than in ab. *lanceolata* Shipp and with the white pupils not extended it is named ab. *cuneata* Gillmer. The ground colour of the upper side may be yellow ochre, ab. *ochracea* Hauder, or pale yellowish brown, ab. *pallens* Schultz. A small undersized specimen has been named ab. *minor* Fuchs. As in other members of this group bleaching on one or more of the wings in the form of irregular patches or blotches is not uncommon and may be referred to as pathological.

Plate 36

Hipparchia semele Linnaeus The Grayling

Although somewhat local in its distribution, this species is found throughout the British Isles wherever the terrain is suitable. The butterfly appears on the wing about the middle of July and flies throughout August into the first half of September. It frequents rough stony ground and bare hillsides in the dry and most sun-baked areas in chalk and sandy districts. As the butterfly always rests with wings closed and with the forewings lowered and concealed behind the hindwings, which match perfectly with the rough ground on which it alights, it presents a most remarkable instance of protective resemblance. It becomes almost invisible when at rest, however carefully the position is marked. In addition to this the butterfly when alighting will run a few inches along the ground then tilt over sideways until it is almost listing flat on the ground, thereby eliminating any shadow thrown by the sun.

There are several subspecies and the most remarkable of these is the dwarf race from the coast of North Wales, this being strikingly smaller than other races or forms from Britain and is referred to as ssp. *thyone* Thompson. The author has met with similar dwarfed specimens in the district of Arnside in Cumbria, where approximately one in four are smaller even than the specimens from North Wales. Those from the chalk districts are generally but not always lighter in colour than the heathland specimens. These have not been separated as a subspecies.

Variation is concerned mostly with the spotting on the upperside of the forewings. With one extra black spot in cell 3 of the forewing it is referred to as ab. *punctata* Aigner. With two extra spots, one between the two normal ones and the other below them, ab. *quadriocellata* Lempke. When the lower of the two spots is either absent or reduced to a mere point it is ab. *monocellata* Lempke. Very rarely the spotting may be absent on all wings, ab. *holanops* Brouwer. With all spotting present but strikingly smaller it is referred to as ab. *parviocellata* Lempke. At the other extreme it may be greatly enlarged and this occurs mostly in the females; it is then named ab. *macrocellata* Lempke. Not infrequently the spot at the anal angle of the hindwings may be absent and is known as ab. *postcaeca* Schawerda. There may be some variation in the ground colour; when a rich orange-fulvous it is ab. *clara* Tutt. The ground colour of the upperside may be paler and of a greyish tint, ab. *grisescens* Lempke. The basal area of the forewing of the female may have a light fulvous patch. An example of this is shown on Plate 36, Fig. 3, and is known as ab. *fulvina* Cabeau. With two extra spots between the two normal ones on the forewing and with two more in addition to the anal spot on the hindwing it is referred to as ab. *addenda* Tutt.

PLATES 37, 38 & 39

Maniola jurtina Linnaeus The Meadow Brown

Although every form of variation is found in this very common species, collectors on the whole appear to give more attention to the spectactular aberrations found in the Fritillaries and the Blues. A careful study can prove most rewarding as there are a great many named forms, the majority of which can be found in our lanes and fields. Three of the coloured plates have been allocated to this species and most of the outstanding aberrations are illustrated.

The butterfly is found all over the British Isles as far north as the Shetlands and the Orkneys. There are several subspecies, the most interesting being the Irish race known as ssp. *iernes* Graves with larger and more extensive orange markings on the upperside, particularly in the females. The subspecies from the Scilly Isles is similar but somewhat smaller. The butterfly can be met with everywhere, in fields, woods, lanes and by roadsides. It will remain in flight in all kinds of weather. When disturbed it has a rapid but erratic flight so that birds give it scant attention. The author has on occasion captured specimens with the imprint of a bird's beak across the wings where the vital parts have been missed, possibly, as believed in some circles, the apical "eye-spot" acts as a decoy. The butterfly is in flight over a long period, the first males appearing in the wing about the middle of June. In suitable seasons it will still be flying as late as October. Although reputed to be single brooded it seems certain that when conditions are right a second brood occurs.

Amongst the many named colour forms a dark and almost black male without any orange marking is named ab. *suffusa* Tutt. A similar blackish form of the female with the fulvous area almost absent and the only colouring a narrow orange circle surrounding the apical spot is ab. *nigrianira* Johnstone. This is shown on Plate 39, Fig. 3. The fulvous patch on the upperside of the forewings of the female may be reddish or dark, ab. *antirufa* Leeds, or whitish yellow, ab. *pallens* Thiery-Mieg, or completely white, ab. *alba* Blackie. Another accepted name for this white form is ab. *antialba* Leeds. A female with the ground colour almost white which may be either an albino or one of the many pale colour forms found in this species has been given the name ab. *cinerea* Cosmovici. A silvery-grey form is also albinistic and is referred to as ab. *grisea-argentacea* Oberthür. The name ab. *excessa* Leeds covers a wide range of extra spotting on all wings, but with extra spotting on the forewings other than those of a double apical spot and similar to ab. *excessa* Leeds of *Pyronia tithonus* is referred to as ab. *addenda* Mosley. The apical spot on the forewings may vary in size. When very large it is named ab. *anticrassipuncta* Leeds, smaller than normal, ab. *antiparvipuncta* Leeds. These are fairly common but a very rare aberration with the spot completely absent occurs occasionally and is named ab. *anommata* Verity. Specimens of this aberration in both sexes are illustrated. The next three variations in the apical spot are met with fairly frequently. With two white pupils, ab. *bioculata* Rebel. With the apical spot divided into two separate spots, ab. *biocellata* Lempke. With the apical "blind" or having no white pupil, ab. *caeca* Rebel. Another accepted name for this form is ab. *pupilatanulla* Leeds. There is a beautiful underside aberration of the female named ab. *postaurolancea* Leeds where the dark central area extends across the median band in long rays or wedges due to the veins being dusted with dark scales. Sometimes there is only one such ray dividing the median band into two sections, ab. *fracta* Zweigelt. These two aberrations are rare but occasionally a colony in a certain field will produce this form at regular intervals over the years.

There are many rayed forms with all the wings streaked horizontally with whitish rays, ab. *radiata* Frohawk. With some or all the wings blotched with whitish it comes under *partimtransformis* Leeds. Homeosis is very rare but specimens have been recorded where the forewing markings have been reproduced on the hindwings, including a somewhat distorted version of the apical spot. A fine example of this aberration was captured by Mr H. G. M. Middleton at Worth, Dorset, in July 1967.

A similar aberration but in *Pyronia tithonus* was taken by the author and is illustrated.

PLATE 40

Pyronia tithonus Linnaeus The Gatekeeper

An alternative English name for this brightly coloured species is the Hedge Brown. In view of its habit of never being found far from the hedgerows the latter seems the more appropriate name. It is also found in large numbers feasting on bramble blossom in fields and around the borders of woods. It is common throughout England and Wales but does not occur in Scotland. In Ireland it is scarce and local, being confined to the southern counties. It appears on the wing in July and when freshly emerged the rich orange colouring of the male is most attractive. The female is more quietly coloured, being paler and more yellowish.

There are no features of special interest about this species but there are a large number of named aberrations. The more important of these are as follows: A completely white ground colour is named ab. *albidus* Cockerell. A pale yellowish-white form is ab. *subalbida* Verity; this form has also been named by Frohawk as ab. *pallidus*. A specimen with the ground colour white on the forewings but normal on the hindwings is ab. *antialba* Leeds. With the outer half of the forewings yellowish but the rest normal, ab. *antipallidula* Leeds. An albino specimen has been recorded with the ground colour of the normal orange but with the darker colouring including the borders replaced by silvery grey and named ab. *albinotica* Goodson. With the ground colour darker than normal and the borders correspondingly darkened referred to as ab. *ultrafulvescens* Leeds. With the forewings normal but with the hindwings so darkened that little or no trace of the orange patch remains it is ab. *postobscura* Leeds.

The remaining aberrations to be described concern the spotting. The apical spot may be enlarged and trailing downwards and is named ab. *anticrassipuncta* Leeds. At the other extreme it may be reduced in size, ab. *antiparvipuncta* Leeds. When exceedingly small and with little or no trace of the white pupil it is referred to as ab. *caeca* Tutt. On the upperside of the forewings there may be anything from one to four extra spots below the apical spot; all these variations are classed under the name ab. *excessa* Tutt. On the upperside of the hindwings there may be three extra spots above the anal spot, ab. *posttriexcessa* Leeds; with four extra spots, ab. *postquadriexcessa* Leeds. Finally with the underside of the hindwings of a uniform darkish brown with no trace of the lighter markings referred to as ab. *infra-unicolora* Lempke.

Plate 40

Coenonympha pamphilus Linnaeus The Small Heath

The Small Heath is found everywhere in the British Isles except in the Shetlands and the Orkneys. It is common everywhere and is as much at home at sea level as it is on the mountains. It has a slow fluttering flight always near the ground. In dull weather it rests on flower-heads and the heads of grasses and other plants. It is in flight during most of the summer, there being two broods. The first appears in May and is on the wing until July; the partial second brood flies during August and September.

There is a great deal of variation but in such a small quietly coloured species nothing spectacular can be expected. Plate 40 shows eight well-contrasted aberrations. Firstly to deal with the ground colour which is very variable and the following shades have been named. The palest form has the colouring whitish yellow and is referred to as ab. *albescens* Robson and Gardner. Next the colouring may be ochreus or very pale orange when it is named ab. *pallidula* Leeds. Specimens showing a reddish tinge or brown with a tinge of red are known as ab. *rufa* Leeds. A dull brownish colouring is ab. *brunnescens* Leeds. Very rarely a purplish suffusion occurs when it is known as ab. *rubescens* Lowe. A dark smoky-grey form has been named by Leeds ab. *obscura*. The centre of both fore and hindwings may be white and this has been named ab. *alba* Pruffer. Another form which occurs not uncommonly in other members of this family has bleached or whitish areas on all wings and been classed under *partimtransformis* by Leeds. A specimen with these whitish markings in a symmetrical pattern is illustrated but more often their placing is irregular.

The remainder of the named aberrations concerns the markings and the most noticeable of these are the variations in the apical spot, particularly on the underside where it may be enlarged, referred to as ab. *crassipuncta* Leeds or reduced ab. *parvipuncta* Leeds. A specimen where it is missing altogether is illustrated and named ab. *caeca* Oberthur. The name *ocellata* has been given by Tutt to a striking aberration of the underside of the hindwings which has a row of strongly marked brownish-red spots, each containing a white pupil. This aberration is also illustrated; so is another hindwing aberration but on the upperside, where the dark border extends broadly inwards. This is more often found in the male and is named ab. *latiora* Leeds. A row of small black spots may appear just inside the border of the hindwings. This has been named by Leeds *postexcessa*. Another form has the blackish borders of all wings pale grey instead of blackish and is known as ab. *latecana* Verity.

PLATE 40

Coenonympha tullia Muller The Large Heath

A close study of this species reveals many interesting features and can prove most rewarding. Its distribution in the British Isles consists of a number of subspecies or geographical forms; these occur in the mosses of the north of England, throughout most of Scotland, parts of North Wales and fairly widespread in Ireland. In western districts the butterfly first appears on the wing about the middle of June but near the east coast and in Scotland it is at least a week later. It has a slow but purposeful flight and is rarely seen to settle, so that a determined chase is necessary before a capture is made for examination, one eye being kept on the water-logged terrain normally selected for its breeding ground.

Three of the local races have been named and each has its distinguishing characteristics. The most striking of these is the race from the mosses of Cumbria known as subspecies *davus* Fabr., with its dark chestnut-brown ground colour and heavy ringed markings on the underside of the hind-wings. In strong contrast the Scottish race is much paler and the ocellated rings are reduced to a minimum. Fig. 20 shows a specimen where all the rings are absent and has been named ab. *obsoleta* Tutt. The Scottish race is referred to as subspecies *scotica* Staudinger. The most widespread is the subspecies *polydama* Haworth. This form is generally intermediate between the two races already described and is found in the mosses of northern England, southern Scotland, North Wales and Ireland. Specimens may be taken in most colonies showing traces of both the other races, so that the English subspecies may be somewhat ill-defined.

Variation concerns the ground colour and the spotting on the underside which may be either enlarged or reduced. When enlarged (usually seen in the ssp. *davus* Fabr.) it has been named ab. *macrocellata* Lempke. It can also be elongated when it is known as ab. *lanceolata* Arkle, and some striking examples can be found in ssp. *davus* where the spotting is normally prominent (see Figs. 16, 17 and 18). The female is normally paler than the male but when more so than usual it is referred to as ab. *pallida* Osthelder.

The most southerly race is found in Shropshire and belongs to the subspecies *davus* Fabricius. The interesting and handsome aberration named ab. *cockaynei* Hopkins occurs in this colony and has the underside of the hindwings with the white markings unusually extensive, the white median band extends inwards in the form of white streaks breaking up the dark central area. The author has found traces of this aberration in Cumbria.

THE COLOURED PLATES

When a life-study is placed at the side of a set specimen this shows upper and underside of the same specimen.

When the figures are placed down the centre of the plate upper and underside of the same specimen are shown with the set specimen above and a life-study below.

When all the specimens figured are set as cabinet specimens they are numbered across.

PLATE 1. *Hesperiidae* The Skippers

Fig. 1 *Carterocephalus palaemon* Pallas. The Chequered Skipper
♂ ab. *melicertes* Schultz. Northants. 18.5.1948. G. E. Hyde

Fig. 2 *Thymelicus sylvestris* Poda. The Small Skipper
♂ ab. *pallida* Tutt. Portland, Dorset. 24.7.1976. H. G. M. Middleton

Fig. 3 *Thymelicus acteon* Rottemburg. The Lulworth Skipper
♀ ab. *alba* Bolton. Swanage. 8.8.1950. E. L. Bolton (B.M. N.H., Watson coll.)

Fig. 4 *Hesperia comma* Linnaeus. The Silver-spotted Skipper
♂ ab. *dupuyi* Oberthür. Box Hill. 2.8.1951. A. D. A. Russwurm

Fig. 5 *Ochlodes venata* Bremer & Grey. The Large Skipper
♂ ab. *obscura* Tutt. New Forest. 24.7.1951. A. D. A. Russwurm

Fig. 6 *Pyrgus malvae* Linnaeus. The Grizzled Skipper
♀ ab. *taras* Bergstrasser. Frimley, Surrey. 18.5.1952. A. D. A. Russwurm

1
1a
2a
2
3
3a
4a
4
5
5a
6a
6

PLATE 2

Papilio machaon Linnaeus. The Swallowtail

Fig. 1 ♂ ab. nov. Bred Norfolk Broads. 17.6.1936. L. W. Newman. ex Marcon coll. (B.M. N.H., N. A. Watkins coll.)

Fig. 2 ♂ ab. *obscura* Frohawk. Bred at Chelmsford, larva from Norfolk Broads. ex L. W. Newman. 24.4.1930. H. C. Ferrier. ex H. J. Turner coll. (Smart coll., Saruman Museum)

1

1a

2

PLATE 3

Colias croceus Geoffroy. The Clouded Yellow

Fig. 1 ♂ ab. *nigrofasciata* Verity. Fairoaks, Hants. 9.10.1949. N. I. Watson (B.M. N.H., Watson coll.)

Fig. 2 ♀ ab. *hyerensis* Strand. Swanage. 23.9.1960. R. Hayward (B.M. N.H., Watson coll.)

Fig. 3 ♀ form *helice* Hubner+*pseudomas* Cockerell. Folkestone. 1877. W. J. Austin (A. S. Harmer coll.)

1

1a

2a

2

3

3a

PLATE 4

Gonepteryx rhamni Linnaeus. The Brimstone

Fig. 1 Gynandromorph. Bred from ova. Hunts. July 1928. G. E. Hyde

Fig. 2 ♂ ab. *nigrescens* Hechler. Bred at Surbiton from larva taken on East Coast. 1920. M. Summers (Smart coll., Saruman Museum)

Fig. 3 ♂ ab. *rubescens* Gillmer. Dorking. 1.8.1937. Dr H. King. ex Bolton Coll. (B.M. N.H., Watson coll.)

1a

1

2

2a

3a

3

PLATE 5

Leptidea sinapis Linnaeus. The Wood White

Fig. 1 ♂ upper and underside. ab. *brunneomaculata* Stauder. Northants. 2.6.1965. P. J. Rogers

Fig. 2 ♂ upper and underside. ssp. *juvernica* Williams. Ballyvaughan, Co. Clair. 21.5.1971. P. J. Rogers

Pieris brassicae Linnaeus. The Large White

Figs. 3 & 4 ♂ & ♀ upper and underside. ab. *coerulea* Gardiner. Bred ex C. F. Rivers stock. July 1975 (B.M. N.H., Watson coll.)

Pieris rapae Linnaeus. The Small White

Fig. 5 ♂ upper and underside. ab. *immaculata* de Selys. Porthcurno, Cornwall. 27.4.1970. A. D. A. Russwurm

1

2

3

4

5

6

PLATE 6

Pieris napi Linnaeus. The Green-veined White

Fig. 1 ♂ ab. *sulphurea* Schoy +ab. *fasciata* Kautz. Bred Donegal. July 1947. H. W. Head (B.M. N.H., Watson coll.)

Fig. 2 ♀ ab. *sulphurea* Schoy +ab. *fasciata* Kautz. Bred Donegal. July 1947. H. W. Head (B.M. N.H., Watson coll.)

Fig. 3 ♀ ab. *sulphurea* Schoy +ab. *fasciata* Kautz+ab. *pallidus* Frohawk. Bred J. A. Thompson. 8.7.1951 (B.M. N.H., Watson coll.)

Fig. 4 ♀ ab. *continua* Bryk+ab. *flavicans* Muller. Caithness. 15.7.1938. N. A. Watkins (B.M. N.H., Watson coll.)

Fig. 5 ♀ ab. *flavicans* Muller. Bred Donegal origin. July 1948. N. A. Watkins (B.M. N.H., Watson coll.)

Fig. 6 ♂ ab. *obsoleta* Rober. Bred Wilts. 28.5.1975. R. E. Stockley (B.M. N.H., Watson coll.)

Anthocharis cardamines Linnaeus. The Orange-tip

Fig. 7 ♂ ab. *buschmanni* Muller. Chiddingfold. 24.4.1960. D. R. Rowberry (B.M. N.H., Watson coll.)

Fig. 8 ♂ ab. *ochracea* Salzl. Hythe. 22.8.1938. J. D. & B. W. Bowdler (B.M. N.H., Watson coll.)

Fig. 9 gynandromorph. East Kent. April 1936. H. D. Smart. (B.M. N.H., Watson coll.)

Fig. 10 gynandromorph. Picton, Chester. 11.5.1950. S. G. Smith (B.M. N.H., Bessemer coll.)

1
2
3
4
5
6
7
8
9
10

PLATE 7

Callophrys rubi Linnaeus. The Green Hairstreak

Fig. 1 ♂ ab. nov. West Moors, Dorset. June 1949. I. G. Farwell

Fig. 2 ♂ ab. *caecus* Geoffroy. Wrotham, Kent. May 1950. A. D. A. Russwurm

Strymonidia w-album Knoch. The White-letter Hairstreak

Fig. 3 ♀ ab. *albovirgata* Tutt. Tilshead, Wilts. 22.7.1956. C. G. Lipscomb

Quercusia quercus Linnaeus. The Purple Hairstreak

Fig. 4 ♂ ab. *violacia* Niepelt. Bred New Forest. July 1965. H. G. M. Middleton

Fig. 5 ♀ ab. *caerulescens* Lempke. Bred New Forest. July 1965. H. G. M. Middleton

Fig. 6 ♀ ab. *flavimaculatus* Lienard (=*bellus* Gerhardt). Ramberry Woods. L. W. Scott (B.M., N.H.)

Fig. 7 ♀ ab. *latefasciata* Courv. Bred Ashtead Wood, Surrey. July 1964. S. F. Imber

Fig. 8 ♂ ab. *infraobscura* Goodson. Bred Ashtead Wood, Surrey. July 1964. S. F. Imber

Thecla betulae Linnaeus. The Brown Hairstreak

Fig. 9 ♂ ab. *steinbuhleri* Hoffman. Bred New Forest. 1921. B. S. Williams (B.M. N.H.)

Fig. 10 ♀ ab. *fisonii* Wheeler. Bred New Forest. 22.7.1899. L. Anning (B.M. N.H., Watson coll.)

Fig. 11 ♀ ab. *unistrigata* Schultz. Bred Winchester. 15.8.1961 (B.M. N.H., R.C.K. coll.)

Lycaena phlaeas Linnaeus. The Small Copper

Fig. 12 ♂ ab. *fuscae* Robson. Boldre, Hants. 3.8.1976. H. G. M. Middleton

Fig. 13 ♂ ab. *berviniensis* Smith. Wembury Point, Devon. 10.10.1969. A. P. Gainsford (B.M. N.H.)

Fig. 14 ♂ ab. *parvipuncta-obsoleta* Tutt. Boldre, Hants. 1.8.1976. A. D. A. Russwurm

Fig. 15 ♀ *bipunctata* Tutt. 4.10.1933. J. Shepherd (B.M. N.H., Watson coll.)

Fig. 16 ♀ ab. *centriconjuncta* Tutt. Romsey. 26.9.1947. A. L. Ransome (B.M. N.H., Watson coll.)

Fig. 17 ♂ ab. *antiradiata* Derenne. Sussex. 25.7.1949 (B.M. N.H., Watson coll.)

Fig. 18 ♂ ab. *cuprinus* Peyer+*parvipuncta* Tutt. Prosperous, Co. Kildare. 31.5.1934 (B.M. N.H., R.C.K. coll.)

Fig. 19 ♀ ab. *coeruleo-fasciata* Holmes. Bred Derbyshire. 1944. Dr J. W. O. Holmes

Fig. 20 ♀ scale defect. Chorley Wood. 1.8.1965. T. S. Robertson (B.M. N.H., Watson coll.)

1
2
3
4
5
6
7
8
9
10
11
12
13
14
15
16
17
18
19
20

PLATE 8

Celastrina argiolas Linnaeus. The Holly Blue

Fig. 1 ♂ ab. *grisescens* Eisner. Homington Down, Wilts. 1952. R. Garner (B.M. N.H., Watson coll.)

Fig. 2 ♀ ab. *aquilina* Grund. Pamber Forest. 8.7.1976. R. L. Harvey

Fig. 3 ♀ ab. *subtus-radiata* Oberthür. Albury, Surrey. 27.7.1944. ex J. C. B. Craske coll. (B.M. N.H.)

Plebejus argus Linnaeus. The Silver-studded Blue.

Fig. 4 ♂ ssp. *argus* ab. *lilacina* Tutt. Worth, Dorset. July 1951. R. W. Watson (B.M. N.H., Watson coll.)

Fig. 5 ♂ ssp. *argus* ab. nov. Bournemouth. 11.7.1937 (B.M. N.H., Watson coll.)

Fig. 6 ssp. *argus* gynandromorph. New Forest. 26.6.1927. M. C. Reid (B.M. N.H., Watson coll.)

Fig. 7 ssp. *argus* intersex. Crowthorne, Berks. July 1925. C. Wells (B.M. N.H., Watson coll.)

Fig. 8 ♂ ssp. *cretaceus* Tutt. ab. *juncta* Tutt. Nodland. 8.7.1915 (B.M. N.H., Watson coll.)

Fig. 9 ♂ ssp. *cretaceus* Tutt. ab. *juncta* Tutt. Maidstone, Kent. 1924. W. Crocker (B.M. N.H., Watson coll.)

Fig. 10 ♂ ssp. *argus* ab. *caeca* Grund. Beaulieu, New Forest. 4.7.1947. A. L. Ransome (B.M. N.H., Watson coll.)

Fig. 11 ♀ ssp. *argus* ab. *privata* Tutt. 9.7.1971. H. G. M. Middleton

Fig. 12 ♀ ssp. *argus* ab. *juncta* Tutt. New Forest. 13.7.1964. A. D. A. Russwurm

Fig. 13 ♀ ssp. *argus* ab. *juncta* Tutt. New Forest. July 1960. N. I. Watson (B.M. N.H., Watson coll.)

Fig. 14 ♀ ssp. *argus* ab. *costajuncta-bibasijuncta* Tutt. Worth, Dorset. 12.7.1951. K. W. Self (B.M. N.H., Watson coll.)

Fig. 15 ♀ ssp. *argus* ab. *juscescens* Lempke. New Forest. 8.7.1954. A. D. A. Russwurm

Aricia agestis Denis & Schiffermüller. The Brown Argus

Fig. 16 ♂ ab. *graafi* Ver-Huell. Stockbridge. 9.8.1976. T. W. Tolman

Fig. 17 ♀ ab. *pallidior* Oberthür. Herne Bay, Kent. August 1897. B. Wallace (B.M. N.H., Watson coll.)

Fig. 18 ♀ ab. *subtus-radiata* Oberthür. Chipstead, Surrey. 8.9.1918. T. L. Barnett (B.M. N.H., Watson coll.)

Fig. 19 ♀ ab. *subtus-radiata* Oberthür. Corfe, Dorset. 1.5.1948. K. W. Self (B.M. N.H., Watson coll.)

Fig. 20 ♀ ab. *deleta* Cockerell. Halnaker Gallop, West Sussex. August 1926. H. G. (B.M. N.H., Watson coll.)

Aricia artaxerxes Fabricius. The Northern Brown Argus

Fig. 21 ♂ ab. *quadripuncta* Tutt. Kincardineshire. July 1915. G. E. Hartley (B.M. N.H., Watson coll.)

Fig. 22 ♂ ssp. *salmacis* Stephens. ab. *unicolor* Lempke. Smardale, Cumbria. 4.8.1972. A. D. A. Russwurm

Fig. 23 ♀ ssp. *salmacis* Stephens. ab. *obsoleta* Tutt. Smardale, Cumbria. 4.8.1972. H. G. M. Middleton

Hamearis lucina Linnaeus. The Duke of Burgundy Fritillary

Fig. 24 ♂ ab. *leucodes* Lamb. Kimble, Bucks. 29.5.1950. A. L. Goodson (B.M. N.H.)

Fig. 25 ♀ ab. *semibrunnea* Osth. Blean Wood, Kent. 31.5.1931. J. Shepherd (B.M. N.H.)

Fig. 26 ab. nov. albino. Portsmouth. June 1919. A. H. Sperring (B.M. N.H.)

1
2
3
4
5
6
7
8
9
10
11
12
13
14
15
16
17
18
19
20
21
22
23
24
25
26

PLATE 9

Polyommatus icarus Rottemburg. The Common Blue

Fig. 1 ♂ ab. *livida* Gillmer. Hampshire Downs. 2.6.1968. R. E. Stockley (B.M. N.H., Watson coll.)

Fig. 2 bilateral gynandromorph. Banstead, Surrey. 31.7.1948. E. L. Bolton (B.M. N.H., Watson coll.)

Fig. 3 ssp. *mariscolore* Kane. gynandromorph. Co. Galway. 24.7.1927. L. A. Sabine. Smart coll. (Saruman Museum)

Fig. 4 ssp. *mariscolore* Kane. gynandromorph. Sligo. 2.7.1952. Rev. J. N. Marcon. Smart coll. (Saruman Museum)

Fig. 5 ♂ ab. *radiata* Courv. Tollard Royal, Wilts. 14.6.1947 (B.N. N.H., Watson coll.)

Fig. 6 ♀ ab. *rufina* Oberthur. Ashbury. June 1942. B. H. Smith (B.M. N.H., Watson coll.)

Fig. 7 ♂ ab. *radiata* Courv. Dorking. 2.6.1934. J. Tetley (B.M. N.H., Watson coll.)

Fig. 8 ♂ ab. *radiata* Courv. South Yorks. 17.6.1944. G. E. Hyde

Fig. 9 ♂ ab. *albescens* Tutt.+ab. *obsoleta* Gillmer. Corfe, Dorset. 5.8.1947 (B.M. N.H., Watson coll.)

Fig. 10 ♂ ssp. *mariscolore* Kane. ab. *postico-obsoleta* Tutt. Co. Tyrone. 8.7.1927. T. Suer (T. G. Farwell coll.)

1
2
3
4
5
6
7
8
9
10

PLATE 10

Lysandra coridon Poda. The Chalk-hill Blue

Fig. 1 ♂ ab. *cinnameus* B. & L. Fovant, Wilts. August 1956. R. W. Watson (B.M. N.H., Watson coll.)

Fig. 2 ♀ ab. *cinnameus* B. & L.+ab. *semisyngrapha* Tutt. Brighton District. August 1911. C. W. P. (B.N. N.H., Watson coll.)

Fig. 3 ♂ ab. *plumbescens* Tutt. extreme form. Folkestone. 29.7.1953 (B.M. N.H., Watson coll.)

Fig. 4 ♀ ab. *tithonus* Meigen=ab. *syngrapha* Keferst. Chilterns, Bucks. August 1917. G. B. Oliver (A. D. A. Russwurm coll.)

Fig. 5 ♂ ab. *ultrafowleri-margino* B. &. L. Portland, Dorset. 12.7.1975. A. D. A. Russwurm

Fig. 6 ♂ ab. *fowleri* South=ab. *grisea* Tutt. Corfe, Dorset. 23.8.1967. A. D. A. Russwurm

Fig. 7. ♀ ab. *semisyngrapha* Tutt+ab. *fowleri* South. Bred F2 generation. July 1976. R. C. Revels

Fig. 8 ♀ ab. *tithonus* Meigen+ab. *fowleri* South. Bred F7 of tithonus ♀. F2 of ♂. 5.8.1974. R. S. Tubbs

Fig. 9 gynandromorph. Shawford, Hants. 9.8.1951. R. W. Watson (B.M. N.H., Watson coll.)

Fig. 10 ♀ ab. *inaequalis* Tutt. Portland, Dorset. July 1974. A. S. Harmer

1

2

3

4

5

6

7

8

9

10

PLATE 11

Lysandra coridon Poda. The Chalk-hill Blue

Fig. 1 ♂ ab. *alba* B & L. South Hants. 3.8.1951. A. L. Ransome (B.M. N.H., Watson coll.)

Fig. 2 ♂ ab. *radiata* B & L. Worth, Dorset. 30.7.1951. K. W. Self (B.M. N.H., Watson coll.)

Fig. 3 ♂ ab. *confluens* B & L. Corfe, Dorset. 1.8.1946. K. W. Self (B.M. N.H., Watson coll.)

Fig. 4 ♂ ab. *radiata* B & L. King's Somborne, Hants. 24.7.1976. N. B. Potter (L. D. Young coll.)

Fig. 5 ♂ ab. *extrema* B & L. Corfe, Dorset. 26.7.1946. K. W. Self (B.M. N.H., Watson coll.)

Fig. 6 ♂ ab. *extrema* B & L. South Hants. 26.7.1950. A. L. Ransome (B.M. N.H., Watson coll.)

Fig. 7 ♀ ab. *antiextrema* B & L.+ab. *postalba* B & L.+*limbojuncta* Courv. Shoreham, Sussex. 6.8.1955. R. E. Stockley (B.M. N.H., Watson coll.)

Fig. 8 ♀ ab. *antiextrema* B & L.+ab. *postalba* B & L.+ab. *limbojuncta* Courv. I.o.W. 15.8.1967. J. F. Johnstone (B.M. N.H., Watson coll.)

Fig. 9 ♀ ab. *alba* B & L.+ab. *anticaeca* B & L.+*prostradiata* B & L. Shoreham, Sussex. 12.8.1957. E. L. Bolton (B.N. N.H., Watson coll.)

Fig. 10 ♀ ab. *ultraradiata* B & L. Portland, Dorset. 29.7.1975. A. D. A. Russwurm

1

2

3

4

5

6

7

8

9

10

PLATE 12

Lysandra bellargus Rottemburg. The Adonis Blue

Fig. 1 ♂ ab. *suffusa* Tutt. Folkestone. September 1916. G. B. Oliver (C. G. Lipscomb coll.)

Fig. 2 ♂ ab. *totonigra* Lipscomb. Hod Hill. 10.6.1965. C. G. Lipscomb

Fig. 3 bilateral gynandromorph. Folkestone. 29.8.1932. Rev. J. N. Marcon (H. J. Turner coll.)

Fig. 4 mixed gynandromorph. Folkestone. 29.8.1931. H. Wood (H. J. Turner coll.)

Fig. 5 ♀ ab. *semiceronus* Tutt. Corfe Castle. 6.6.1968. A. D. A. Russwurm

Fig. 6 ♀ ab. *caerulescens-lunulata* Tutt. Corfe Castle. 12.6.1968 A. D. A. Russwurm

Fig. 7 ♂ ab. *krodeli* Gillmer. Hod Hill. 8.9.1961. E. L. Bolton (B.M. N.H., Watson coll.)

Fig. 8 ♂ ab. *krodeli* Gillmer. Hod Hill. September 1959. A. D. A. Russwurm

Fig. 9 ♂ peroneural form. Chilterns. September 1943. Dr A. Harris (H. J. Turner coll.)

Fig. 10 ♂ ab. *conjuncta* Tutt. Folkestone. September 1927. L. W. Newman (B.M. N.H., Watson coll.)

1
2
3
4
5
6
7
8
9
10

PLATE 13

Maculinea arion Linnaeus. The Large Blue

Fig. 1 ♂ ab. *alconides* Aurivillius. Bude, Cornwall. 30.6.1905. L. A. Sabine (A. S. Harmer coll.)

Fig. 2 ♂ ab. *arthurus* Melvill. Bude, Cornwall. 15.7.1905 (A. J. Turner coll.)

Fig. 3 ♀ ab. *pallida* le Chamberlin+ab. *insubrica* Vorbr. Bude, Cornwall. July 1902. A. Druitt (A. S. Harmer coll.)

Fig. 4 ♀ ab. *conjuncta* Tutt. North Cornwall. 3.7.1899 (A. S. Harmer coll.)

Fig. 5 ♀ ab. *marginata* le Chamberlin. Dartmoor. 8.7.1937 (B.M. N.H., Watson coll.)

Fig. 6 ♀ ab. *magnifica* Heyd. North Cornwall. July 1902 (A. S. Harmer coll.)

Fig. 7 ♂ ab. *supra-impuncta* Oberthür+ab. *marginata* le Chamberlin. Bude, Cornwall. 20.8.1902 (B.M. N.H., Watson coll.)

Fig. 8 ♂ ab. *supra-impuncta* Oberthur. Bude, Cornwall. 20.8.1902 (B.M. N.H., Watson coll.)

1

2

3

4

5

6

7

8

PLATE 14

Ladoga camilla Linnaeus. The White Admiral

Fig. 1 ♀ ab. *obliterae* Robson & Gardner. New Forest. 26.6.1956. A. D. A. Russwurm

Fig. 2 ♂ ab. *nigrina* Weymer. Cranbourne Chase. 19.6.1976. R. E. Stockley (B.M. N.H., Watson coll.)

Fig. 3 ♂ ab. *nigrina* Weymer. New Forest. July 1956. R. W. Watson (B.M. N.H., Watson coll.)

1

1a

2a

2

3

3a

PLATE 15

Apatura iris Linnaeus. The Purple Emperor

Figs. 1 & 1a ♂ upper & underside ab. *chattendeni* Heslop & Stockley. Chattenden. July 1878. T. W. Wood. ex Bolton coll. (B.M. N.H., Watson coll.)

Fig. 2 ♂ ab. *monophana* Cabeau. Bred from larva. West Sussex. 10.8.1938. H.G. (B.M. N.H., Watson coll.)

1

1a

2

PLATE 16

Apatura iris Linnaeus. The Purple Emperor

♀ ab. *iolata* Cabeau (=*semi-iole* Frohawk). Captured Alice Holt Wood. 28.7.1965. R. C. Revels

1

1a

PLATE 17

Aglais urticae Linnaeus. The Small Tortoiseshell

Fig. 1 ♂ ab. *dannenbergi* Neuburg, extreme form. Bexley, Kent. 9.8.1945. R. L. E. Ford. ex J. C. B. Craske coll. (B.M. N.H., Watson coll.)

Fig. 2 ♀ ab. *lutea* Raynor. Bred New Forest. 22.6.1972. A. D. A. Russwurm

Fig. 3 ♀ ab. *lutea* Raynor+ab. *semiichnusoides* Pronin. Chilterns, Bucks. September 1918. G. B. Oliver. ex. E. L. Bolton coll. (B.M. N.H., Watson coll.)

1

1a

2a

2

3

3a

PLATE 18

Aglais urticae Linnaeus. The Small Tortoiseshell

Fig. 1 ♂ ab. *semiichnusoides* Pronin. New Forest. 21.7.1975. A. D. A. Russwurm

Fig. 2 ♀ ab. *conjuncta* Meuberg. Ringwood, Hants. 30.9.1947. D. Watson. ex E. L. Bolton coll. (B.M. N.H., Watson coll.)

1

2

PLATE 19

Vanessa cardui Linnaeus. The Painted Lady

Fig. 1 ♂ ab. *varini* Meilhan. Colchester. 1917. H. J. Turner coll. (Smart coll., Saruman Museum)

Vanessa atalanta Linnaeus. The Red Admiral

Fig. 2 ♀ ab. *klemensiewiczi* Schille. Poynings, Sussex. 29.7.1971. R. M. Craske (B.M. N.H.)

Inachis io Linnaeus. The Peacock

Fig. 3 ♀ ab. *prochnovi* Pronin. Bred Coleshill. 20.7.1943. G. P. Sutton (B.M. N.H., Watson coll.)

1

1a

2a

2

3

3a

PLATE 20

Nymphalis polychloros Linnaeus. The Large Tortoiseshell

Fig. 1 ♂ ab. *testudo* Esper. Suffolk. A. M. Adams (B.M. N.H., Watson coll.)

Fig. 2 ♀ ab. *testudo* Esper. Bred New Forest, 1902. Mr Simmons (B.M. N.H., Watson coll.)

1

1a

2

PLATE 21

Polygonia c-album Linnaeus. The Comma

Fig. 1 ♂ ab. *reichstettensis* Fettig. Captured Vernditch Wood. 10.7.1976 (C. G. Lipscomb coll.)

Fig. 2 ♂ ab. *reichstettensis* Fettig. Bred Forest of Dean. July 1934. G. B. Oliver (B.M. N.H., Watson coll.)

Fig. 3 ♂ ab. *obscura* Closs. Captured New Forest. 22.9.1964. A. D. A. Russwurm

1

1a

2a

2

3

3a

PLATE 22

Boloria (*Clossiana*) *selene* Denis & Schiffermuller. The Small Pearl-bordered Fritillary

Fig. 1 ♂ ab. *pallida* Spuler. Aberdeenshire. 16.7.1907. G. Hartley (B.M. N.H., Watson coll.)

Fig. 2 ♂ ab. nov. Bred Chiddingfold. June 1970. R. E. Stockley (B.M. N.H., Watson coll.)

Fig. 3 ♂ ab. *vanescens* Cabeau. New Forest. July 1931 (A. S. Harmer coll.)

Fig. 4 ♂ ab. *e-nigrum* Foltin. Worth, Sussex. 18.5.1945. J. Tetley (B.M. N.H., Watson coll.)

Fig. 5 ♂ ab. *nigricans* Oberthur. Polgate, Sussex. 1.6.1930 (A. S. Harmer coll.)

1

1a

2a

2

3

3a

4a

4

5

5a

PLATE 23

Boloria (*Clossiana*) *euphrosyne* Linnaeus. The Pearl-bordered Fritillary

Fig. 1 ♂ ab. *tatrica* Aigner. New Forest. 14.5.1971. A. D. A. Russwurm

Fig. 2 ♂ ab. *pittionii* Nitsche. Hampshire Downs. 11.5.1967. M. V. Bolton (B.M. N.H., Watson coll.)

Fig. 3 ♀ ab. *nigricans* Oberthür. Midhurst (B.M. N.H., Watson coll.)

Fig. 4 ♀ ab. nov. Cotswolds. 20.5.1929. E. D. F. (B.M. N.H., Watson coll.)

1

1a

2a

2

3

3a

4a

4

PLATE 24

Argynnis (*Fabriciana*) *adippe* Linnaeus. The High Brown Fritillary

Fig. 1 ♂ ab. *margareta* Stephan. Haresfield. 9.7.1939. ex L. Birch coll. (B.M. N.H., Watson coll.)

Fig. 2 ♀ ab. *margareta* Stephan. Three Bridges. August 1877. J. P. Barrett (B.M. N.H., Watson coll.)

Fig. 3 ♀ ab. *bronzus* Frohawk. Wyre Forest. 14.7.1941. G. P. Sutton (B.M. N.H., Watson coll.)

1

1a

2a

2

3

3a

PLATE 25

Argynnis (*Mesoacidalia*) *aglaja* Linnaeus. The Dark Green Fritillary

Fig. 1 ♂ upper & underside ab. *nigrans* Newnham. Swinley Woods, Berks. 10.7.1962. R. E. Stockley (B.M. N.H., Watson coll.)

Fig. 2 ♀ upper & underside ab. *wimani* Holmgren. Nr Canterbury. 1906. ex Bessemer coll. (B.M. N.H.)

1

2

PLATE 26

Argynnis paphia Linnaeus. The Silver-washed Fritillary

Fig. 1 Mixed gynandromorph. New Forest. 6.7.1942. J. Tetley (H. J. Turner coll.)

Fig. 2 ♂ ab. *ocellata* Frings. Wilts. 6.7.1976. R. E. Stockley (H. Phelps coll.)

Fig. 3 ♂ ab. *confluens* Spuler. Wilts. 5.7.1976. R. E. Stockley (M. Hale coll.)

1a

1

2

2a

3a

3

PLATE 27

Argynnis paphia Linnaeus. The Silver-washed Fritillary

Fig. 1 ♀ ab. *ocellata* Frings. Wilts. 9.7.1976. H. Phelps

Fig. 2 ♀ ab. *nigricans* Cosmovici. Wilts. 7.7.1976. C. G. Lipscomb

Fig. 3 ♀ ab. *valesina* Esper+ab. *ocellata* Frings. Wilts. 10.7.1976. H. Phelps

1

1a

2a

2

3

3a

PLATE 28

Argynnis paphia Linnaeus. The Silver-washed Fritillary

Fig. 1 ♀ ab. nov. Bred New Forest from *valesina*. ♀ 26.6.1975. R. C. Revels

Fig. 2 ♀ ab. *ocellata* Frings. Wilts. 7.7.1976. R. E. Stockley (M. Hale coll.)

1

1a

2

PLATE 29

Euphydryas aurinia Rottemburg. The Marsh Fritillary

Fig. 1 ♂ ab. *melanoleuca* Cabeau. Bratton, Wilts. 22.5.1971. R. E. Stockley (B.M. N.H., Watson coll.)

Fig. 2 ♂ ab. *atricolor* Schultz. Hod Hill, Dorset. 29.5.1963. C. G. Lipscomb

Fig. 3 ♀ ab. *epimolpadia* Rev. Hod Hill, Dorset. 1.6.1966. A. D. A. Russwurm

Fig. 4 ♂ ab. *melanoleuca* Cabeau. Hod Hill, Dorset. 7.6.1975. A. S. Harmer

1a

1

2

2a

3a

3

4

4a

PLATE 30

Melitaea cinxia Linnaeus. The Glanville Fritillary

Fig. 1 ♂ ab. *horvathi* Aigner. Bred Portmore, Hants. ex Brook, I.o.W. 21.5.1961. I. G. Farwell

Fig. 2 ♂ ab. nov. Bred Brockenhurst, ex I.o.W. 19.6.1971. A. D. A. Russwurm

Fig. 3 ♂ ab. *wittei* Geest. Bred Boldre, Hants. 4.6.1975. H. G. M. Middleton

Fig. 4 ♀ ab. *horvathi* Aigner. Bred I.o.W. 7.6.1929. G. Nobbs (H. J. Turner coll.)

1a

1

2

2a

3a

3

4

4a

PLATE 31

Mellicta athalia Rottemburg. The Heath Fritillary

Fig. 1 ♀ ab. *corythallia* Hubner. Church Wood, Kent. 24.6.1973. A. S. Harmer (A. S. Harmer coll.)

Fig. 2 ♀ ab. *latonigena* Spuler. East Kent. 7.6.1935. G. B. Oliver (B.M. N.H., Watson coll.)

Fig. 3 ♀ ab. *cymothoe* Bertolini. Blean, Kent. 29.6.1973. C. J. Luckens (C. J. Luckens coll.)

Fig. 4 ♀ ab. *cymothoe* Bertolini. Blean, Kent. 29.6.1950. E. L. Bolton, ex E. L. Bolton coll. (B.M. N.H., Watson coll.)

1
1a
2a
2
3
3a
4a
4

PLATE 32

Pararge aegeria Linnaeus. The Speckled Wood

Fig. 1 ♂ ab. *intermediana* Lempke. Boldre, Hants. 17.4.1967.
A. D. A. Russwurm

Fig. 2 ♂ ab. *cockaynei* Goodson. Paignton. 5.11.1932. P. P. Milman (B.M. N.H., Watson coll.)

Fig. 3 ♀ ab. nov. enlarged spotting. New Forest. 5.8.1966.
A. D. A. Russwurm

Fig. 4 ♂ ab. *saturiatior* Crombrugghe. Kimbolton, Hunts. 15.7.1944.
A. D. A. Russwurm

1

1a

2a

2

3

3a

4a

4

PLATE 33

Lasiommata megera Linnaeus. The Wall

Fig. 1 ♂ ab. *boltoni* Watson. Shoreham, Sussex. 31.7.1953. J. F. Johnstone (B.M. N.H., Watson coll.)

Fig. 2 ♀ ab. *melania* Oberthur. Orlestone. 26.5.1960 (B.M. N.H., Watson coll.)

Fig. 3 ♀ ab. *fascia* Frohawk. Sanderstead, Surrey. 8.8.1948. R. E. Stockley (B.M. N.H. Watson coll.)

Fig. 4 ♀ ab. *castaneopicta* Verity. Sidcup. 21.7.1949 (B.M. N.H., Watson coll.)

Fig. 5 ♂ pathological. Surrey. 29.5.1954 (B.M. N.H., Watson coll.)

Fig. 6 ♀ ab. *croesus* Stauder=ab. *triopes* Muschamp=ab. *limbojuncta* Stammehaus. Portsmouth. 19.8.1968. J. F. Johnstone (B.M. N.H., Watson coll.)

Fig. 7 ♀ pathological. Royston, Herts. 8.8.1952. A. D. A. Russwurm

Fig. 8 ♀ ab. *xanthos* Frohawk. Lollington Hill, Berks. 25.8.1968. R. E. Stockley (B.M. N.H., Watson coll.)

1

2

3

4

5

6

7

8

PLATE 34

Erebia epiphron Knoch. The Small Mountain Ringlet.

Fig. 1 ♂ ab. *nelamus* Boisduval. Ben Lawers, Perthshire. 3.7.1973. A. D. A. Russwurm

Fig. 2 ♂ ab. *caeca* Vorbrodt. Buttermere Fell, Cumbria. 24.6.1969. A. D. A. Russwurm

Fig. 3 ♀ ab. *albinescens* Oberthür. Ben Lawers, Perthshire. 9.7.1963. P. J. Rogers

Fig. 4 ♂ ab. *obsoleta* Tutt. Ben Lawers, Perthshire. 30.6.1973. R. W. Watson (B.M. N.H., Watson coll.)

Fig. 5 ♀ ab. *privata* Turati & Verity. Ben Lawers, Perthshire. 7.7.1973. H. G. M. Middleton

Fig. 6 ♀ ab. *latefasciata* Dioszeghy. Ben Lawers, Perthshire. 7.7.1973. H. G. M. Middleton

Erebia aethiops Esper. The Scotch Argus

Fig. 7 ♂ enlarged spotting. Arnside, Cumbria. 2.8.1971. A. D. A. Russwurm

Fig. 8 ♂ ab. *infasciata* Warren. Arnside, Cumbria. 1.8.1971. A. D. A. Russwurm

Fig. 9 ♂ ab. *perfumosa* Dannehl. Arnside, Cumbria. 3.8.1971. A. D. A. Russwurm

Fig. 10 ♀ ab. *leucotaenia* Staudinger. Smardale, Cumbria. 10.8.1972. A. D. A. Russwurm

Fig. 11 ♀ ab. *infasciata* Warren. Arnside, Cumbria. 3.8.1971. H. G. M. Middleton

Fig. 12 ♀ enlarged and elongated spotting. Arnside, Cumbria. 1.8.1971. A. D. A. Russwurm

Figs. 13 & 14. bilateral gynandromorph, upper & underside. Arnside, Cumbria. 4.8.1970. A. D. A. Russwurm

Figs. 15 & 16. upperside bilateral gynandromorph, underside mixed gynandromorph. ab. *ochracea* Mosley 5.8.1974. C. R. Haxby (B.M. N.H., Watson coll.)

1
2
3
4
5
6
7
8
9
10
11
12
13
14
15
16

PLATE 35

Melanargia galathea Linnaeus. The Marbled White

Fig. 1 ♂ ab. *grisescens* Varin. Charterhouse, Somerset. 26.7.1967. R. W. Watson (B.M. N.H., Watson coll.)

Fig. 2 ♂ ab. *mosleyi* Oberthur. Bred Sussex. 1953. A. E. Collier (C. G. Lipscomb coll.)

Fig. 3 ♂ ab. *valentini* Williams. Somerset. 7.7.1950. Arthur Valentine (B.M. N.H., Watson coll.)

Fig. 4 ♂ ab. *galene* Ochsenheimer. Wilnecote, Staffs. 6.7.1918. C. W. B. J. (B.M. N.H., Watson coll.)

Aphantopus hyperantus Linnaeus. The Ringlet

Fig. 5 ♂ ab. *arete* Muller+ab. *lanceolata* Shipp. Bred Crockerton, Wilts. July 1976. C. G. Lipscomb

Fig. 6 ♀ ab. *arete* Muller+ab. *lanceolata* Shipp. Bred Crockerton, Wilts. July 1976. C. G. Lipscomb

Fig. 7 ♂ ab. *lanceolata* Shipp. New Forest. July 1958. A. D. A. Russwurm

Fig. 8 ♀ ab. *lanceolata* Shipp. Fritham, New Forest. July 1958. C. R. Haxby (B.M. N.H., Watson coll.)

Fig. 9 ♀ ab. *chrysopharis* Collier. New Forest. 14.7.1967. A. D. A. Russwurm

Fig. 10 ♂ ab. nov. New Forest. 13.7.1964. A. D. A. Russwurm

1

2

3

4

5

6

7

8

9

10

PLATE 36

Hipparchia semele Linnaeus. The Grayling

Fig. 1 ♂ ab. *holonops* Brouwer. Portland, Dorset. 5.8.1975. D. B. Tyler

Fig. 2 ♂ ab. *suffusa* Tutt. Portland, Dorset. 8.8.1977. A. D. A. Russwurm

Fig. 3 ♀ ab. *fulvina* Cabeau. Portland, Dorset. 21.8.1965. R. W. Watson (B.M. N.H., Watson coll.)

Fig. 4 ♀ ab. *monocellata* Lempke. Portland, Dorset. 29.7.1975. A. D. A. Russwurm

1

1a

2a

2

3

3a

4a

4

PLATE 37

Maniola jurtina Linnaeus. The Meadow Brown

Fig. 1 ♂ transformis. Brendor, Devon. 21.7.1973. I. G. Farwell

Fig. 2 ♂ ab. *radiata* Frohawk. Worth, Dorset. 15.8.1949. F. & L. Britton. (B.M. N.H., Watson coll.)

Fig. 3 ♂ ab. *atrescens* Leeds. Worth, Dorset. 26.7.1969. H. G. M. Middleton

Fig. 4 ♂ ab. *anommata* Verity. New Forest. 28.7.1963. A. D. A. Russwurm

Fig. 5 ♂ ab. *subtus-albida* Silbernagel. Eynsford, Kent. 2.8.1947. A. D. A. Russwurm

Fig. 6 ♂ ab. *cinerea* Cosmovici. Worth, Dorset. 27.7.1950. K. W. Self (B.M. N.H., Watson coll.)

Figs. 7 & 8 ♂ upper and underside. ab. *excessa* Leeds. Devonshire. 16.7.1925. K. W. Self (B.M. N.H., Watson coll.)

1
2
3
4
5
6
7
8

PLATE 38

Maniola jurtina Linnaeus. The Meadow Brown

Fig. 1 ♀ ab. *radiata* Frohawk. Portland, Dorset. 24.6.1976. H. G. M. Middleton

Fig. 2 ♀ ab. nov. Whiteparish, Wilts. 10.7.1973. R. C. Revels

Fig. 3 .b. *anommata* Verity + ab. *antialba* Leeds. Portland, Dorset. 23.8.1972. R. W. Watson (B.M. N.H., Watson coll.)

Fig. 4 ♀ *transformis* Leeds. Portland, Dorset. 26.7.1976. A. D. A. Russwurm

1

1a

2a

2

3

3a

4a

4

PLATE 39

Maniola jurtina Linnaeus. The Meadow Brown

Fig. 1 ♀ ab. *cinerea* Cosmovici. Hod Hill, Dorset. 25.8.1972. R. W. Watson (B.M. N.H., Watson coll.)

Fig. 2 ♀ ab. *nuragiformis* Verity. Portland, Dorset. 6.8.1977. H. G. M. Middleton

Fig. 3 ♀ ab. *nigrianira* Johnstone. Hod Hill, Dorset. 24.8.1974. A. D. A. Russwurm

Fig. 4 ♀ ab. *atrescens* Leeds. Portland, Dorset. 9.8.1970. R. B. Watson (B.M. N.H., Watson coll.)

Fig. 5 ♀ ab. *cinerascens* Fuchs. Portland, Dorset. 24.7.1976. A. D. A. Russwurm

Fig. 6 ♀ ab. *postaurolancea* Leeds. New Forest. 9.7.1969. A. D. A. Russwurm

Fig. 7 ♀ ab. *postaurolancea* Leeds. Mere, Wilts. 17.9.1972. C. G. Lipscomb

Fig. 8 ♀ ab. *postaurolancea* Leeds. Mere, Wilts. 20.9.1972. C. G. Lipscomb

1

2

3

4

5

6

7

8

PLATE 40

Pyronia tithonus Linnaeus. The Gatekeeper

Fig. 1 ♀ ab. *subalbida* Verity. Bernwood Forest. 27.8.1972. D. B. Tyler

Figs. 2 & 5 ♀ upper and underside. ab. *excessa* Tutt+ab. *postobscura* Leeds. Tavistock, Devon. August 1915. A. T. Stiff (B.M. N.H., Watson coll.)

Fig. 3 ♂ homoeosis on right hindwing (underside). New Forest. 28.7.1966. A. D. A. Russwurm

Fig. 4 ♀ ab. *excessa-lanceolata* Frohawk. Boldre, Hants. 23.8.1977. H. G. M. Middleton

Fig. 6 ♀ transitional to ab. *infra-unicolora* Lempke. Boldre, Hants. 24.8.1977. A. D. A. Russwurm

Coenonympha pamphilus Linnaeus. The Small Heath

Fig. 7 ♂ *partimtransformis* Leeds. Rough Tor, Cornwall. August 1937 (B.M. N.H., Watson coll.)

Fig. 8 ♀ ab. *alba* Pruffer. Worth, Dorset. 7.8.1951. K. W. Self (B.M. N.H., Watson coll.)

Fig. 9 ♂ ab. *ocellata* Tutt. Bracon Hill, Hants. 5.6.1965. R. E. Stockley (B.M. N.H., Watson coll.)

Fig. 10 ♀ ab. *aurea* Leeds. Portland, Dorset. 26.7.1976. A. D. A. Russwurm

Fig. 11 ♂ ab. *latiora* Leeds. Suffolk. 15.6.1943 (B.M. N.H., Watson coll.)

Fig. 12 ♂ ab. *excessa* Leeds. Charterhouse, Somerset. 8.7.1966. A. D. A. Russwurm

Fig. 13 ♀ ab. nov. Stockbridge, Hants. August 1966. R. E. Stockley (B.M. N.H., Watson coll.)

Fig. 14 ♂ ab. *caeca* Oberthür. Burley, Hants. 22.8.1976. R. W. Rowley

Coenonympha tullia Muller. The Large Heath

Fig. 15 ♂ ssp. *polydama* Haworth. ab. nov. Goathland, Yorks. July 1951. A. D. A. Russwurm

Figs. 16 & 17 ♂ ssp. *davus* Fabr. ab. *lanceolata* Arkle. Meathop Moss, Westmorland. 6.7.1968. H. G. M. Middleton

Fig. 18 ♀ ssp. *davus* Fabr. ab. *lanceolata* Arkle.+ab. *impupillata* Lempke. Meathop Moss, Westmorland. 6.7.1968. A. D. A. Russwurm

Fig. 19 ♀ ssp. *davus* Fabr. ab. *impupillata* Lempke. Meathop Moss, Westmorland. 3.7.1969. A. D. A. Russwurm

Fig. 20 ♀ ssp. *scotica* Staudinger. ab. *obsoleta* Tutt, extreme form. Ben Lawers, Perthshire. 8.7.1973. A. D. A. Russwurm

1
2
3
4
5
6
7
8
9
10
11
12
13
14
15
16
17
18
19
20

INDEX